Structural Systems

ARE Mock Exam

(SS of Architect Registration Exam)

ARE Overview, Exam Prep Tips,
Multiple-Choice Questions and Graphic Vignettes,
Solutions and Explanations

Gang Chen

ArchiteG®, Inc.
Irvine, California

Structural Systems ARE Mock Exam (SS of Architect Registration Exam): ARE Overview, Exam Prep Tips, Multiple-Choice Questions and Graphic Vignettes, Solutions and Explanations

Copyright © 2013	Gang Chen
V1.8	Incorporated minor revisions on 1/2/2015
Cover Photo © 2013	Gang Chen

Copy Editor: Penny L Kortje

ArchiteG®, Inc.
http://www.ArchiteG.com

ISBN: 978-1-61265-001-2

PRINTED IN THE UNITED STATES OF AMERICA

Dedication

To my parents, Zhuixian and Yugen,
my wife, Xiaojie, and my daughters,
Alice, Angela, Amy, and Athena.

Disclaimer

Structural Systems ARE Mock Exam (SS of Architect Registration Exam) provides general information about Architect Registration Exam. The book is sold with the understanding that neither the publisher nor the authors are providing legal, accounting, or other professional services. If legal, accounting, or other professional services are required, seek the assistance of a competent professional firm.

The purpose of this publication is not to reprint the content of all other available texts on the subject. You are urged to read other materials, and tailor them to fit your needs.

Great effort has been taken to make this resource as complete and accurate as possible. However, nobody is perfect and there may be typographical errors or other mistakes present. You should use this book as a general guide and not as the ultimate source on this subject. If you find any potential errors, please send an e-mail to:
info@ArchiteG.com

Structural Systems ARE Mock Exam (SS of Architect Registration Exam) is intended to provide general, entertaining, informative, educational, and enlightening content. Neither the publisher nor the author shall be liable to anyone or any entity for any loss or damages, or alleged loss or damages, caused directly or indirectly by the content of this book.

If you do not wish to be bound by the above, you may return this book to the publisher for a full refund.

Legal Notice

ARE Mock Exam series by ArchiteG, Inc.

Time and effort is the most valuable asset of a candidate. How to cherish and effectively use your limited time and effort is the key of passing any exam. That is why we publish the ARE Mock Exam series to help you to study and pass the ARE exams in the shortest time possible. We have done the hard work so that you can save time and money. We do not want to make you work harder than you have to.

Do not force yourself to memorize a lot of numbers. Read through the numbers a few times, and you should have a very good impression of them.

You need to make the judgment call: If you miss a few numbers, you can still pass the exam, but if you spend too much time drilling these numbers, you may miss out on the big pictures and fail the exam.

The official NCARB sample MC questions have no explanations, and the official NCARB sample vignettes have no step-by-step solutions. The existing ARE practice questions or exams by others are either way too easy or way over-killed. They do NOT match the real ARE exams at all.

We have done very comprehensive research on the official NCARB guides, many related websites, reference materials, and other available ARE exam prep materials. We match our mock exams vignettes questions as close as possible to the NCARB samples and the real ARE exams instead. Some other readers had failed an ARE exam two or three times before, and they eventually passed the exam with our help.

All our books include a step-by-step solution to each of the NCARB sample vignette with screenshots and related NCARB software commands, a complete set of MC questions (except for the SD ARE mock exam) and vignettes matching the real ARE exams, including number of questions, format, type of questions, etc. We also include detailed answers and explanations to our MC questions, and a step-by-step solution to each of the mock vignette with screenshots and related NCARB software commands. Our DWG files for our mock exam vignettes can also be installed and used with the NCARB software.

There is some extra information on ARE overviews and exam-taking tips in Chapter One. This is based on NCARB AND other valuable sources. This is a bonus feature we included in each book because we want our readers to be able to buy our ARE mock exam books together or individually. We want you to find all necessary ARE exam information and resources at one place and through our books.

All our books are available at
http://www.GreenExamEducation.com

How to Use This Book

We suggest you read *Structural Systems ARE Mock Exam (SS of Architect Registration Exam)* at least three times:

Read once and cover Chapter One and Two, the Appendixes, the related FREE PDF files, and other resources. Highlight the information you are not familiar with.

Read twice focusing on the highlighted information to memorize. You can repeat this process as many times as you want until you master the content of the book.

After reviewing these materials, you can take the mock exam, and then check your answers against the answers and explanations in the back, including explanations for the questions you answer correctly. You may have answered some questions correctly for the wrong reason. Highlight the information you are not familiar with.

Like the real exam, the mock exam includes three question types: Select the correct answer, check all that apply, and fill in the blank.

Review your highlighted information, and take the mock exam again. Try to answer 100% of the questions correctly this time. Repeat the process until you can answer all the questions correctly.

SS is one of the most difficult ARE divisions because many SS questions require calculations. This book includes most if not all the information you need to do the calculations, as well as step-by-step explanations. After reading this book, you will greatly improve your ability to deal with the real ARE SS calculations, and have a great chance of passing the exam on the first try.

Take the mock exam at least two weeks before the real exam. You should definitely NOT wait until the night before the real exam to take the mock exam. If you do not do well, you will go into panic mode and NOT have enough time to review your weaknesses.

Read for the final time the night before the real exam. Review ONLY the information you highlighted, especially the questions you did not answer correctly when you took the mock exam for the first time.

One important tip to pass the graphic vignette section of the ARE SS division is to become VERY familiar with the commands of the NCARB software. Many people fail the exam simply because they are NOT familiar with the NCARB software and cannot finish the graphic vignette section within the time limit.

For the graphic vignette section, we included a step-by-step solution using screen-shots from the NCARB practice program so that you can use the book to become familiar with the commands, even when you do NOT have a computer in front of you. This book is very light so you can easily carry it around. These features will allow you to review the graphic vignette section whenever you have a few minutes.

All commands are described in an **abbreviated manner**. For example, **Draw > Column** means go to the menu on the left hand side of your computer screen, click **Draw,** and then click **Column** to draw a column**.** This is typical for ALL commands throughout the book.

The Table of Contents is very detailed so you can locate information quickly. If you are on a tight schedule you can forgo reading the book linearly and jump around to the sections you need.

All our books, including "ARE Mock Exams Series" and "LEED Exam Guides Series," are available at
GreenExamEducation.com

Check out FREE tips and info at **GeeForum.com**, you can post your questions or vignettes for other users' review and responses.

Table of Contents

Chapter Four   ARE Mock Exam Solutions for Structural Systems (SS) Division

Appendixes

Back Page Promotion
A. **ARE Mock exam series (GreenExamEducation.com)**
B. **LEED Exam Guides series (GreenExamEducation.com)**
C. *Building Construction* **(ArchiteG.com)**
D. *Planting Design Illustrated*

Index

Chapter One

Overview of the Architect Registration Exam (ARE)

A. First Thing First: Go to the Website of your Architect Registration Board and Read all the Requirements of Obtaining an Architect License in your Jurisdiction
See the following link:
http://www.ncarb.org/Getting-an-Initial-License/Registration-Board-Requirements.aspx

B. Download and Review the Latest ARE Documents at the NCARB Website

1. Important links to the FREE and official NCARB documents
The current version of the Architect Registration Exam includes seven divisions:

- Programming, Planning & Practice
- Site Planning & Design
- Building Design & Construction Systems
- Schematic Design
- Structural Systems
- Building Systems
- Construction Documents and Services

Note: Starting July 2010, the 2007 AIA Documents apply to all ARE Exams.

Six ARE divisions have a multiple-choice section and a graphic vignette section. The Schematic Design division has NO multiple-choice section, but two graphic vignette sections.

For the vignette section, you need to complete the following graphic vignette(s) based on the ARE division you are taking:

Programming, Planning & Practice
Site Zoning

Site Planning & Design
Site Grading
Site Design

Building Design & Construction Systems
Accessibility/Ramp
Stair Design
Roof Plan

Schematic Design
Interior Layout
Building Layout

Structural Systems
Structural Layout

Building Systems
Mechanical & Electrical Plan

Construction Documents & Services
Building Section

There is a tremendous amount of valuable information covering every step of becoming an architect available free of charge at the NCARB website:
http://www.ncarb.org/

For example, you can find the education guide regarding professional architectural degree programs accredited by the National Architectural Accrediting Board (NAAB), NCARB's Intern Development Program (IDP) guides, initial license, certification and reciprocity, continuing education, etc. These documents explain how you can qualify to take the Architect Registration Exam.

I find the official ARE Guidelines, exam guide, and practice program for each of the ARE divisions extremely valuable. See the following link:
http://www.ncarb.org/ARE/Preparing-for-the-ARE.aspx

You should definitely start by studying the official exam guide and practice program for the ARE division you are taking.

2. **A detailed list and brief description of the FREE PDF files that you can download from NCARB**
    The following is a detailed list of the FREE PDF files that you can download from NCARB. They are listed in order based on their importance.

    *   **ARE Guidelines** includes extremely valuable information on the ARE overview, six steps to complete ARE, multiple-choice section, graphic vignette section, exam format, scheduling, sample exam computer screens, links to other FREE NCARB PDF files, practice software for graphic vignettes, etc. You need to read this <u>at least twice</u>.

    *   **NCARB Education Guidelines** (Skimming through this should be adequate)

- **Intern Development Program Guidelines** contains important information on IDP overview, IDP steps, IDP reporting, IDP basics, work settings, training requirements, supplementary education (core), supplementary education (elective), core competences, next steps, and appendices. Most of NCARB's 54-member boards have adopted the IDP as a prerequisite for initial architect licensure. This is why you should be familiar with it. IDP costs $350 for the first three years, and then $75 annually. The fees are subject to change, and you need to check the NCARB website for the latest information. Your IDP experience should be reported to NCARB at least every six months and logged within two months of completing each reporting period (the **Six-Month Rule**). You need to read this document at least twice. It has a lot of valuable information.

- **The IDP Supervisor Guidelines** (Skimming through this should be adequate. You should also forward a copy of this PDF file to your IDP supervisor.)

- **Handbook for Interns and Architects** (Skimming through this should be adequate.)

- **Official exam guide, references index, and practice program (NCARB software) for each ARE division**
  This includes specific information for each ARE division. (Just focus on the documents related to the ARE divisions you are currently taking and read them at least twice. Make sure you install the practice program and become very familiar with it. The real exam is VERY similar to the practice program.)

  a. **Programming, Planning & Practice (PPP)**: Official exam guide and practice program for the PPP division
  b. **Site Planning & Design (SPD)**: Official exam guide and practice program (computer software) for the SPD division
  c. **Building Design & Construction Systems (BDCS)**: Official exam guide and practice program for the BDCS division
  d. **Schematic Design (SD)**: Official exam guide and practice program for the SD division
  e. **Structural Systems (SS)**: Official exam guide, references index, and practice program for the SS division
  f. **Building Systems (BS)**: Official exam guide and practice program for the BS division
  g. **Construction Documents and Services (CDS)**: Official exam guide and practice program for the CDS division

- **The Burning Question: Why Do We Need ARE Anyway?** (Skimming through this should be adequate.)

- **Defining Your Moral Compass** (Skimming through this should be adequate.)

- **Rules of Conduct** is available as a FREE PDF file at:
  http://www.ncarb.org/
  (Skimming through this should be adequate.)

C. The Intern Development Program (IDP)

1. What is IDP?

IDP is a comprehensive training program jointly developed by the National Council of Architectural Registration Boards (NCARB) and the American Institute of Architects (AIA) to ensure that interns obtain the necessary skills and knowledge to practice architecture <u>independently</u>.

2. Who qualifies as an intern?

Per NCARB, if an individual meets one of the following criteria, s/he qualifies as an intern:
a. Graduates from NAAB-accredited programs
b. Architecture students who acquire acceptable training prior to graduation
c. Other qualified individuals identified by a registration board

D. Overview of the Architect Registration Exam (ARE)

1. How to qualify for the ARE?

A candidate needs to qualify for the ARE via one of NCARB's member registration boards, or one of the Canadian provincial architectural associations.

Check with your Board of Architecture for specific requirements.

For example, in California, a candidate must provide verification of a minimum of <u>five</u> years of education and/or architectural work experience to qualify for the ARE.

Candidates can satisfy the five-year requirement in a variety of ways:

- Provide verification of a professional degree in architecture through a program that is accredited by NAAB or CACB.

  OR
- Provide verification of at least five years of educational equivalents.

  OR
- Provide proof of work experience under the direct supervision of a licensed architect

2. How to qualify for an architect license?

Again, each jurisdiction has its own requirements. An individual typically needs a combination of about <u>eight</u> years of education and experience, as well as passing scores on the ARE exams. See the following link:

http://www.ncarb.org/Reg-Board-Requirements

For example, the requirements to become a licensed architect in California are:

- Eight years of post-secondary education and/or work experience as evaluated by the Board (including at least one year of work experience under the direct supervision of an architect licensed in a U.S. jurisdiction or two years of work experience under the direct supervision of an architect registered in a Canadian province)
- Completion of the Comprehensive Intern Development Program (CIDP) and the Intern Development Program (IDP)
- Successful completion of the Architect Registration Examination (ARE)
- Successful completion of the California Supplemental Examination (CSE)

California does NOT require an accredited degree in architecture for examination and licensure. However, many other states do.

3. What is the purpose of ARE?

The purpose of ARE is NOT to test a candidate's competency on every aspect of architectural practice. Its purpose is to test a candidate's competency on providing professional services to protect the <u>health, safety, and welfare</u> of the public. It tests candidates on the <u>fundamental</u> knowledge of pre-design, site design, building design, building systems, and construction documents and services.

The ARE tests a candidate's competency as a "specialist" on architectural subjects. It also tests her abilities as a "generalist" to coordinate other consultants' works.

You can download the exam content and references for each of the ARE divisions at the following link:

http://www.ncarb.org/are/40/StudyAids.html

4. What is NCARB's rolling clock?

a. Starting on January 1, 2006, a candidate MUST pass ALL ARE sections within five years. A passing score for an ARE division is only valid for five years, and a candidate has to retake this division if she has NOT passed all divisions within the five year period.

b. Starting on January 1, 2011, a candidate who is authorized to take ARE exams MUST take at least one division of the ARE exams within five years of the authorization. Otherwise, the candidate MUST apply for the authorization to take ARE exams from an NCARB member board again.

These rules were created by the **NCARB's rolling clock** resolution and passed by NCARB council during the 2004 NCARB Annual Meeting.

5. How to register for an ARE exam?

See the instructions in the new ARE guideline at the following link:
http://www.ncarb.org/en/ARE/~/media/Files/PDF/Guidelines/ARE_Guidelines.pdf

6. How early do I need to arrive at the test center?

Be at the test center at least 30 minutes BEFORE your scheduled test time, OR you may lose your exam fee.

7. Exam format & time

All ARE divisions are administered and graded by computer. Their detailed exam format and time allowances are as follows:

1) Programming, Planning & Practice (PPP)

Introduction Time:	15 minutes	
MC Testing Time:	**2 hours**	**85 items**
Scheduled Break:	15 minutes	
Introduction Time:	15 minutes	
Graphic Testing Time:	**1 hour**	**Site Zoning (1 vignette)**
Exit Questionnaire:	15 minutes	
Total Time	**4 hours**	

2) Site Planning & Design (SPD)

Introduction Time:	15 minutes	
MC Testing Time:	**1.5 hours**	**65 items**
Scheduled Break:	15 minutes	
Introduction Time:	15 minutes	
2 Graphic Vignettes:	**2 hours**	**Site Grading, Site Design**
Exit Questionnaire:	15 minutes	
Total Time	**4.5 hours**	

3) Building Design & Construction Systems (BDCS)

Introduction Time:	15 minutes	
MC Testing Time:	**1.75 hours**	**85 items**
Scheduled Break:	15 minutes	
Introduction Time:	15 minutes	
3 Graphic Vignettes:	**2.75 hours**	**Accessibility/Ramp, Stair Design, Roof Plan**
Exit Questionnaire:	15 minutes	
Total Time	**5.5 hours**	

4) Schematic Design (SD)

Introduction Time:	15 minutes	
Graphic Testing Time:	**1 hour**	**Interior Layout (1 vignette)**
Scheduled Break:	15 minutes	
Introduction Time:	15 minutes	
Graphic Testing Time:	**4 hours**	**Building Layout (1 vignette)**
Exit Questionnaire:	15 minutes	
Total Time	**6 hours**	

5) Structural Systems (SS)

Introduction Time:	15 minutes	
MC Testing Time:	**3.5 hours**	**125 items**
Scheduled Break:	15 minutes	
Introduction Time:	15 minutes	
Graphic Testing Time:	**1 hour**	**Structural Layout (1 vignette)**
Exit Questionnaire:	15 minutes	
Total Time	**5.5 hours**	

6) Building Systems (BS)

Introduction Time:	15 minutes	
MC Testing Time:	**2 hours**	**95 items**
Scheduled Break:	15 minutes	
Introduction Time:	15 minutes	
Graphic Testing Time:	**1 hour**	**Mechanical & Electrical Plan (1 vignette)**
Exit Questionnaire:	15 minutes	
Total Time	**4 hours**	

7) Construction Documents and Services (CDS)

Introduction Time:	15 minutes	
MC Testing Time:	**2 hours**	**100 items**
Scheduled Break:	15 minutes	
Introduction Time:	15 minutes	
Graphic Testing Time:	**1 hour**	**Building Section (1 vignette)**
Exit Questionnaire:	15 minutes	
Total Time	**4 hours**	

8. How are ARE scores reported?

All ARE scores are reported as Pass or Fail. ARE scores are typically posted within 5 to 10 days since 2013. See the instructions in the new ARE guideline at the following link: http://www.ncarb.org/en/ARE/~/media/Files/PDF/Guidelines/ARE_Guidelines.pdf

9. Is there a fixed percentage of candidates who pass the ARE exams?

No, there is NOT a fixed percentage of passing or failing. If you meet the minimum competency required to practice as an architect, you pass. The passing scores are the same for all Boards of Architecture.

10. When can I retake a failed ARE division?

You can only take the same ARE division once within a 6-month period.

11. How much time do I need to prepare for each ARE division?

Every person is different, but on average you need about 40 to 80 hours to prepare for each ARE division. You need to set a realistic study schedule and stick with it. Make sure you allow time for personal and recreational commitments. If you are working full time, my suggestion is that you allow no less than 2 weeks but NOT more than 2 months to prepare for each ARE division. You should NOT drag out the exam prep process too long and risk losing your momentum.

12. Which ARE division should I take first?

This is a matter of personal preference, and you should make the final decision.

Some people like to start with the easier divisions and pass them first. This way, they build more confidence as they study and pass each division.

Other people like to start with the more difficult divisions so that if they fail, they can keep busy studying and taking the other divisions while the clock is ticking. Before they know it, six months has passed and they can reschedule if need be.

Programming, Planning & Practice (PPP) and Building Design & Construction Systems (BDCS) divisions often include some content from the Construction Documents and Service (CDS) division. It may be a good idea to start with CDS and then schedule the exams for PPP and BDCS soon after.

13. ARE exam prep and test-taking tips

You can start with Construction Documents and Services (CDS) and Structural Systems (SS) first because both divisions give a limited scope, and you may want to study building regulations and architectural history (especially famous architects and buildings that set the trends at critical turning points) before you take other divisions.

Complete mock exams and practice questions and vignettes, including those provided by NCARB's practice program and this book, to hone your skills.

Form study groups and learn the exam experience of other ARE candidates. The forum at our website is a helpful resource. See the following link:
http://GreenExamEducation.com/

Take the ARE exams as soon as you become eligible, since you probably still remember portions of what you learned in architectural school, especially structural and architectural history. Do not make excuses for yourself and put off the exams.

The following test-taking tips may help you:
- Pace yourself properly. You should spend about one minute for each Multiple-Choice (MC) question, except for the SS division questions which you can spend about one and a half minutes on.
- Read the questions carefully and pay attention to words like *best, could, not, always, never, seldom, may, false, except,* etc.
- For questions that you are not sure of, eliminate the obvious wrong answer and then make an educated guess. Please note that if you do NOT answer the question, you automatically lose the point. If you guess, you at least have a chance to get it right.
- If you have no idea what the correct answer is and cannot eliminate any obvious wrong answers, then do not waste too much time on the question and just guess. Try to use the same guess answer for all of the questions you have no idea about. For example, if you choose "d" as the guess answer, then you should be consistent and use "d" whenever you have no clue. This way, you are likely have a better chance at guessing more answers correctly.
- Mark the difficult questions, answer them, and come back to review them AFTER you finish all MC questions. If you are still not sure, go with your first choice. Your first choice is often the best choice.
- You really need to spend time practicing to become VERY familiar with NCARB's graphic software and know every command well. This is because the ARE graphic vignette is a timed test, and you do NOT have time to think about how to use the software during the test. If you do not know how, you will NOT be able to finish your solution to the vignette on time.
- The ARE exams test a candidate's competency to provide professional services protecting the <u>health, safety, and welfare</u> of the public. Do NOT waste time on aesthetic or other design elements not required by the program.

ARE exams are difficult, but if you study hard and prepare well, combined with your experience, IDP training, and/or college education, you should be able to pass all divisions and eventually be able to call yourself an architect.

14. Strategies for passing ARE exams on the first try
Passing ARE exams on the first try, like everything else, needs not only hard work, but also great strategy.

- **Find out how much you already know and what you should study**
  You goal is NOT to read all the study materials. Your goal is to pass the exam. Flip through the study materials. If you already know the information, skip these parts.

  Complete the NCARB sample vignette(s) and MC questions for the ARE exam you are preparing for NOW without ANY studying. See what percentage you get right. If you get 60% right, you should be able to pass the real exam without any studying. If you get 50% right, then you just need 10% more to pass.

This "truth-finding" exam or exercise will also help you to find out what your weakness areas are, and what to focus on.

Look at the same questions again at the end of your exam prep, and check the differences.

Note: We suggest you study the vignettes in the official NCARB Study Guide first, and then study the official NCARB Multiple Choice (MC) sample questions, and then other study materials, and then come back to NCARB vignette and the NCARB (MC) questions again several days before the real ARE exam.

- **Cherish and effectively use your limited time and effort**

  Let me paraphrase a story.
  One time someone had a chance to talk with Napoleon. He said:
  "You are such a great leader and have won so many battles, that you can use one of your soldiers to defeat ten enemy soldiers."

  Napoleon responded:
  "That may be true, but I always try to create opportunities where ten of my soldiers fight one enemy soldier. That is why I have won so many battles."

  Whether this story is true is irrelevant. The important thing that you need to know is **how to concentrate your limited time and effort to achieve your goal. Do NOT spread yourself too thin**. This is a principle many great leaders know and use and is why great leaders can use ordinary people to achieve extraordinary goals.

  Time and effort is the most valuable asset of a candidate. How to cherish and effectively use your limited time and effort is the key to passing any exam.

  If you study very hard and read many books, you are probably wasting your time. You are much better off picking one or two good books, covering the major framework of your exams, and then doing two sets of mock exams to find your weaknesses. You WILL pass if you follow this advice. You may still have minor weakness, but you will have covered your major bases.

- **Do NOT stretch your exam prep process too long**
  If you do this, it will hurt instead of help you. You may forget the information by the time you take the exam.

  Spend 40 to 80 hours for each division (a maximum of two months for the most difficult exams if you really need more time) should be enough. Once you decide on taking an exam, put in 100% of your effort and read the RIGHT materials. Allocate your time and effort on the most important materials, and you will pass.

- **Nailing the vignette**
  This is the easiest way to improve your chance of passing any ARE exam on the first try, and can be done in a very short period. You really need to spend time practicing to become VERY familiar with NCARB's graphic software. You need to practice it many times until you know it like the back of your hand.

  If you fail any ARE exam because of the vignette(s), you ONLY have yourself to blame.

  Note: Again, we suggest you study the vignettes in the official NCARB Study Guide first, and then study the official NCARB Multiple Choice (MC) sample questions, and then other study materials, and then come back to NCARB vignette and the NCARB (MC) questions again several days before the real ARE exam.

- **Resist the temptation to read too many books and limit your time and effort to read only a few selected books or a few sections of each book in detail**
  Having all the books but not reading them, or digesting ALL the information in them will not help you. It is like someone having a garage full of foods, and not eating or eating too much of them. Neither way will help.

  You can only eat three meals a day. Similarly, you can ONLY absorb a certain amount of information during your exam prep. So, focus on the most important stuff.

  Focus on your weaknesses but still read the other info. The key is to understand, digest the materials, and retain the information.

  It is NOT how much you have read, but how much you understand, digest, and retain that counts.

  The key to passing an ARE exam, or any other exam, is to know the scope of the exam, and not to read too many books. Select one or two really good books and focus on them. Actually <u>understand</u> the content and <u>memorize</u> it. For your convenience, I have <u>underlined</u> the fundamental information that I think is very important. You definitely need to <u>memorize</u> all the information that I have underlined.

  You should try to understand the content first, and then memorize the content of the book by reading it multiple times. This is a much better way than relying on "mechanical" memory without understanding.

When you read the materials, ALWAYS keep the following in mind:

- **Think like an architect.**
  For example, when you take the SS ARE exam, focus on what need to know to be able to coordinate your structural engineer's work, or tell them what to do. You are NOT taking an exam for becoming a structural engineer; you are taking an exam to become an architect.

  This criterion will help you filter out the materials that are irrelevant, and focus on the right information. You will know what to flip through, what to read line by line, and what to read multiple times.

  I have said this one thousand times, and I will say it once more:
  Time and effort is the most valuable asset of a candidate. How to cherish and effectively use your limited time and effort is the key to passing any exam.

15. ARE exam preparation requires short-term memory

You should understand that ARE Exam Preparation requires **Short-Term Memory**. This is especially true for the MC portion of the exam. You should schedule your time accordingly: in the early stages of your ARE exam Preparation, you should focus on understanding and an **initial** review of the material; in the late stages of your exam preparation, you should focus on memorizing the material as a **final** review.

16. Allocation of your time and scheduling

You should spend about 60% of your effort on the most important and fundamental study materials, about 30% of your effort on mock exams, and the remaining 10% on improving your weakest areas, i.e., reading and reviewing the questions that you answered incorrectly, reinforcing the portions that you have a hard time memorizing, etc.

Do NOT spend too much time looking for obscure ARE information because the NCARB will HAVE to test you on the most **common** architectural knowledge and information. At least 80% to 90% of the exam content will have to be the most common, important and fundamental knowledge. The exam writers can word their questions to be tricky or confusing, but they have to limit themselves to the important content; otherwise, their tests will NOT be legally defensible. At most, 10% of their test content can be obscure information. You only need to answer about 60% of all the questions correctly. So, if you master the common ARE knowledge (applicable to 90% of the questions) and use the guess technique for the remaining 10% of the questions on the obscure ARE content, you will do well and pass the exam.

On the other hand, if you focus on the obscure ARE knowledge, you may answer the entire 10% obscure portion of the exam correctly, but only answer half of the remaining 90% of the common ARE knowledge questions correctly, and you will fail the exam. That is why we have seen many smart people who can answer very difficult ARE questions correctly because they are able to look them up and do quality research. However, they often end up failing ARE exams because they cannot memorize the common ARE knowledge needed on the day of the exam. ARE exams are NOT an open-book exams, and you cannot look up information during the exam.

The **process of memorization** is like **filling a cup with a hole at the bottom**: You need to fill it <u>faster</u> than the water leaks out at the bottom, and you need to <u>constantly</u> fill it; otherwise, it will quickly be empty.

Once you memorize something, your brain has already started the process of forgetting it. It is natural. That is how we have enough space left in our brain to remember the really important things.

It is tough to fight against your brain's natural tendency to forget things. Acknowledging this truth and the fact that you ca<u>nnot</u> memorize everything you read, you need to <u>focus</u> your limited time, energy and brainpower on the <u>most important</u> issues.

The biggest danger for most people is that they memorize the information in the early stages of their exam preparation, but forget it before or on the day of the exam and still THINK they remember them.

Most people fail the exam NOT because they cannot answer the few "advanced" questions on the exam, but because they have read the information but can <u>NOT</u> recall it on the day of the exam. They spend too much time preparing for the exam, drag the preparation process on too long, seek too much information, go to too many websites, do too many practice questions and too many mock exams (one or two sets of mock exams can be good for you), and **spread themselves too thin**. They end up **missing the most important information** of the exam, and they will fail.

The ARE Mock Exam series along with the tips and methodology in each of the books will help you find and improvement your weakness areas, MEMORIZE the most important aspects of the test to pass the exam ON THE FIRST TRY.

So, if you have a lot of time to prepare for the ARE exams, you should plan your effort accordingly. You want your ARE knowledge to peak at the time of the exam, not before or after.

For example, <u>if you have two months to prepare for a very difficult ARE exam</u>, you may want to spend the first month focused on <u>reading and understanding</u> all of the study materials you can find as your **initial** review. Also during this first month, you can start <u>memorizing</u> after you understand the materials as long as you know you HAVE to review the materials again later to <u>retain</u> them. If you have memorized something once, it is easier to memorize it again later.

Next, you can spend two weeks focused on <u>memorizing</u> the material. You need to review the material at least three times. You can then spend one week on <u>mock exams</u>. The last week before the exam, focus on retaining your knowledge and reinforcing your weakest areas. Read the mistakes that you have made and think about how to avoid them during the real exam. Set aside a mock exam that you <u>have not taken</u> and take it seven days before test day. This will alert you to your weaknesses and provide direction for the remainder of your studies.

<u>If you have one week to prepare for the exam</u>, you can spend two days reading and understanding the study material, two days repeating and memorizing the material, two days on mock exams, and one day retaining the knowledge and enforcing your weakest areas.

The last one to two weeks before an exam is <u>absolutely</u> critical. You need to have the "do or die" mentality and be ready to study hard to pass the exam on your first try. That is how some people are able to pass an ARE exam with only one week of preparation.

17. **Timing of review: the 3016 rule; memorization methods, tips, suggestions, and mnemonics**

Another important strategy is to review the material in a timely manner. Some people say that the best time to <u>review</u> material is between <u>30 minutes and 16 hours</u> (the **3016** rule) after you read it for the first time. So, if you review the material right after you read it for the first time, the review may not be helpful.

I have personally found this method extremely beneficial. The best way for me to memorize study materials is to review what I learn during the day again in the evening. This, of course, happens to fall within the timing range mentioned above.

Now that you know the **3016** rule, you may want to schedule your review accordingly. For example, you may want to read <u>new</u> study materials in the morning and afternoon, then after dinner do an <u>initial review</u> of what you learned during the day.

OR

If you are working full time, you can read <u>new</u> study materials in the evening or at night and then get up early the next morning to spend one or two hours on an <u>initial review</u> of what you learned the night before.

The <u>initial</u> review and memorization will make your <u>final</u> review and memorization much easier.

Mnemonics is a very good way for you to memorize facts and data that are otherwise very hard to memorize. It is often <u>arbitrary</u> or <u>illogical</u> but it works.

A good mnemonic can help you remember something for a long time or even a lifetime after reading it just once. Without the mnemonics, you may read the same thing many times and still not be able to memorize it.

There are a few common Mnemonics:
1) **Visual** Mnemonics: Link what you want to memorize to a visual image.
2) **Spatial** Mnemonics: link what you want to memorize to a space, and the order of things in it.
3) **Group** Mnemonics: Break up a difficult piece into several smaller and more manageable groups or sets, and memorize the sets and their order. One example is the grouping of the 10 digit phone number into three groups in the U.S. This makes the number much easier to memorize.
4) **Architectural** Mnemonics: A combination of Visual Mnemonics and Spatial Mnemonics and Group Mnemonics.

Imagine you are walking through a building several times, along the same path. You should be able to remember the order of each room. You can then break up the information that you want to remember and link them to several images, and then imagine you hang the images on walls of various rooms. You should be able to easily recall each group in an orderly manner by imagining you are walking through the building again on the same path, and looking at the images hanging on walls of each room. When you look at the images on the wall, you can easily recall the related information.

You can use your home, office or another building that you are familiar with to build an Architectural Mnemonics to help you to organize the things you need to memorize.

5) **Association** Mnemonics: You can associate what you want to memorize with a sentence, a similarly pronounced word, or a place you are familiar with, etc.
6) **Emotion** Mnemonics: Use emotion to fix an image in your memory.
7) **First Letter** Mnemonics: You can use the first letter of what you want to memorize to construct a sentence or acronym. For example, "**Roy G. Biv**" can be use to memorize the order of the 7 colors of the rainbow, it is composed of the first letter of each primary color.

You can use **Association** Mnemonics and memorize them as all the plumbing fixtures for a typical home, PLUS Urinal.

OR
You can use "Water S K U L" (**First Letter** Mnemonics selected from website below) to memorize them:

Water Closets
Shower
Kitchen Sinks
Urinal
Lavatory

18. The importance of good and effective study methods

There is a saying: Give a man a fish, feed him for a day. Teach a man to fish, feed him for a lifetime. I think there is some truth to this. Similarly, it is better to teach someone HOW to study than just give him good study materials. In this book, I give you good study materials to save you time, but more importantly, I want to teach you effective study methods so that you can not only study and pass ARE exams, but also so that you will benefit throughout the rest of your life for anything else you need to study or achieve. For example, I give you samples of mnemonics, but I also teach you the more important thing: HOW to make mnemonics.

Often in the same class, all the students study almost the SAME materials, but there are some students that always manage to stay at the top of the class and get good grades on exams. Why? One very important factor is they have good study methods.

Hard work is important, but it needs to be combined with effective study methods. I think people need to work hard AND work SMART to be successful at their work, career, or anything else they are pursuing.

19. The importance of repetition: read this book <u>at least</u> three times

Repetition is one of the most important tips for learning. That is why I have listed it under a separate title. For example, you should treat this book as part of the core study materials for your ARE exams and you need to read this book <u>at least three times</u> to get all of its benefits:

1) The first time you read it, it is new information. You should focus on understanding and digesting the materials, and also do an <u>initial</u> review with the **3016** rule.
2) The second time you read it, focus on reading the parts <u>I</u> have already highlighted AND <u>you</u> have <u>highlighted</u> (the important parts and the weakest parts for you).
3) The third time, focus on <u>memorizing</u> the information.

Remember the analogy of the <u>memorization process</u> as **filling a cup with a hole on the bottom**?
Do NOT stop reading this book until you pass the real exam.

20. The importance of a routine

A routine is very important for studying. You should try to set up a routine that works for you. First, look at how much time you have to prepare for the exam, and then adjust your current routine to include exam preparation. Once you set up the routine, stick with it.

For example, you can spend from 8:00 a.m. to 12:00 noon, and 1:00 p.m. to 5:00 p.m. on studying new materials, and 7:00 p.m. to 10:00 p.m. to do an initial review of what you learned during the daytime. Then, switch your study content to mock exams, memorization and retention when it gets close to the exam date. This way, you have 11 hours for exam preparation everyday. You can probably pass an ARE exam in one week with this method. Just keep repeating it as a way to <u>retain</u> the architectural knowledge.

OR

You can spend 7:00 p.m. to 10:00 p.m. on studying new materials, and 6:00 a.m. to 7:00 a.m. to do an initial review of what you learned the evening before. This way, you have four hours for exam preparation every day. You can probably pass an ARE exam in two weeks with this preparation schedule.

A routine can help you to memorize important information because it makes it easier for you to concentrate and work with your body clock.

Do NOT become panicked and change your routine as the exam date gets closer. It will not help to change your routine and pull all-nighters right before the exam. In fact, if you pull an all-nighter the night before the exam, you may do much worse than you would have done if you kept your routine.

All-nighters or staying up late are not effective. For example, if you break your routine and stay up one-hour late, you will feel tired the next day. You may even have to sleep a few more hours the next day, adversely affecting your study regimen.

21. The importance of short, frequent breaks and physical exercise
Short, frequent breaks and physical exercise are VERY important for you, especially when you are spending a lot of time studying. They help relax your body and mind, making it much easier for you to concentrate when you study. They make you more efficient.

Take a five-minute break, such as a walk, at least once every one to two hours. Do at least 30 minutes of physical exercise every day.

If you feel tired and cannot concentrate, stop, go outside, and take a five-minute walk. You will feel much better when you come back.

You need your body and brain to work well to be effective with your studying. Take good care of them. You need them to be well-maintained and in excellent condition. You need to be able to count on them when you need them.

If you do not feel like studying, maybe you can start a little bit on your studies. Just casually read a few pages. Very soon, your body and mind will warm up and you will get into study mode.

Find a room where you will NOT be disturbed when you study. A good study environment is essential for concentration.

22. A strong vision and a clear goal

You need to have a strong vision and a clear goal: to <u>master</u> the architectural knowledge and <u>become an architect in the shortest time</u>. This is your number one priority. You need to master the architectural knowledge BEFORE you do sample questions or mock exams, except "truth-finding" exam or exercise at the very beginning of your exam prep. It will make the process much easier. Everything we discuss is to help you achieve this goal.

As I have mentioned on many occasions, and I say it one more time here because it is so important:

It is how much architectural knowledge and information you can <u>understand, digest, memorize</u>, and firmly retain that matters, not how many books you read or how many sample tests you have taken. The books and sample tests will NOT help you if you cannot understand, digest, memorize, and retain the important information for the ARE exam.

Cherish your limited <u>time and effort</u> and focus on the most <u>important</u> information.

23. English system (English or inch-pound units) vs. metric system (SI units)

This book is based on the English system or English units. The English or inch-pound units are based on the module used in the U.S. Effective July 2013, the ARE includes measurements in inch-pound units only. Metric system (SI units) is no longer used.

24. Codes and standards used in this book

We use the following codes and standards:

American Institute of Architects, Contract Documents, Washington, DC.

Canadian Construction Documents Committee, CCDC Standard Documents, 2006, Ottawa.

25. Where can I find study materials on architectural history?

Every ARE exam may have a few questions related to architectural history. The following are some helpful links to FREE study materials on the topic:

http://www.essentialhumanities.net/arch.php
http://issuu.com/motimar/docs/history_synopsis?viewMode=magazine
http://www.ironwarrior.org/ARE/Materials_Methods/m_m_notes_2.pdf

Chapter Two

Structural Systems (SS) Division

A. General Information

1. Exam content

The exam content for the SS division of the ARE includes:
1) general structural principles (38% to 42% of scored items)
2) seismic forces (28% to 32% of scored items)
3) wind forces (14% to 27% of scored items)
4) lateral forces – general (13% to 16% of scored items)

The exam content can be further broken down as follows:
- the design and construction of buildings
- principles (building design; the implications of design decisions; building systems and their integration)
- materials and technology (construction materials; construction details and constructability)
- codes and regulations (government and regulatory requirements; permit processes)
- concrete, steel, and wood structures
- foundation systems
- connections
- famous engineers and structures
- statics and strength of materials
- structural and truss analysis
- retaining walls and other types of walls

For the graphic vignette, you will be required to design a schematic framing plan for a one-story building. This building has a multi-level roof.

2. Official exam guide, references index, and practice program for the Structural Systems (SS) division

You need to read the official exam guide for the SS division at least twice. Make sure you install the SS division practice program and become very familiar with it. The real exam is VERY similar to the practice program.

You can download the official exam guide, references index, and practice program for the **Structural Systems (SS)** division at the following link:
http://www.ncarb.org/en/ARE/Preparing-for-the-ARE.aspx

Every July, NCARB updates the exam guide and scope.

Note: Again, we suggest you study the vignettes in the official NCARB Study Guide first, and then study the official NCARB Multiple Choice (MC) sample questions, and then other study materials, and then come back to NCARB vignette and the NCARB (MC) questions again several days before the real ARE exam.

B. The Most Important Documents/Publications for SS Division of the ARE Exam

1. Official NCARB list of formulas and references for the Structural Systems (SS) division with our comments and suggestions
You can download the NCARB list of formulas and references for the Structural Systems (SS) division at the following link:
http://www.ncarb.org/en/ARE/Preparing-for-the-ARE.aspx

These formulas and references will be available during the real exam. You should read through them a few times before the exam to become familiar with the content. This will save you a lot of time during the real exam, and will help you solve structural calculations and other problems.

Note:
*While the majority of the MC questions in the real SS ARE exam **focus on structural design concepts**, there are **at least 20 questions requiring calculations**. Therefore, it is absolutely necessary and critical for you to be very familiar with some of the basic and important equations, and to memorize them if possible. We have incorporated some of the most important equations into our SS mock exam.*

In the ARE exams, it may be a good idea to skip any calculation question that requires over 30 seconds of your time; just pick a guess answer, mark it, and come back to calculate it at the end. This way, you have more time to read and answer other easier questions correctly.

A calculation question that takes 20 minutes to answer will gain the same number of points as a simple question that ONLY takes 2 minutes.

If you spend 20 minutes on a calculation question earlier, you risk losing the time to read and answer ten other easier questions, which could result in a loss of ten points instead of one.

The NCARB list of formulas and references includes the following:

- ***Manual of Steel Construction: Allowable Stress Design***; 9th Edition. American Institute of Steel Construction, Inc. Chicago, Illinois, 1989

  Look through the following pages and become familiar with the structural shapes, designations, dimensions, and properties:
  pg. 1-9 thru 1-16, pg. 1-18 thru 1-32, pg. 1-40 thru 1-41, pg. 1-46 thru 1-52, pg. 1-94 thru 1-103.

  Make sure you understand what the designations stand for. For example, on pages 1-10 and 1-11, **W 40 x 298** means a W shape steel with a nominal depth of 40" (the actual depth is 39.69" per the Table on page 1-10), and a nominal weight of 298 lb. per ft. Use the diagram and Tables on pages 1-10 and 1-11 to look up the other detailed properties of a **W 40 x 298.** You do NOT need to remember any of these properties. You just need to know how to look them up and what they mean.

  The most important information for an architect is the overall dimension of a structural member so that you can coordinate and make sure you have enough space to accommodate it. For example, you may need to find out if it will fit inside a wall or interstitial space.

  You can also use the size of a structural member, the mechanical duct size, and the clearance space needed for a light fixture and/or fire sprinkler line to determine the interstitial height between floors.

  Look through the following pages and become familiar with the beam nomenclature, diagrams, and formulas:
  pg. 2-293 & 2-294, pg. 2-296, pg. 2-297, pg. 2-298, pg. 2-301, pg. 2-304, pg. 2-305.

  Look through the following pages and become familiar with bolts, threaded parts, and rivet tensions:
  pg. 4-3 & 4-5.

- ***Steel Construction Manual***; 13th Edition. American Institute of Steel Construction, Inc. Chicago, Illinois, 2005
  Read the following pages and become familiar with the round HSS dimensions and properties:
  pg. 1-94 thru 1-98.

- ***2006 International Building Code (IBC)***. International Code Council, Inc. Country Club Hills, Illinois, 2006

  Focus on **Chapter 16**, particularly the sections on earthquakes and wind. Read them a few times, and have a general idea of the concepts. Do not force yourself to memorize all the details.

Read the following pages and become familiar with the Uniform and Concentrated Loads IBC table 1607.1:

pg. 285 & 286.
OR
You can read the Uniform and Concentrated Loads IBC Table 1607.1 for FREE at the following link:
http://publiccodes.cyberregs.com/icod/ibc/2006f2/icod_ibc_2006f2_16_sec001.htm

Note:

- *The latest version of IBC is the 2012 version, but, in the 2012 NCARB SS Exam Guide, NCARB still list the 2006 version as the official version to use for SS ARE exam.*

- *If the codes are updated by NCARB in the future, you just need to go to the root directory and find the latest version of the codes:*
  http://publiccodes.cyberregs.com/icod/ibc/index.htm

- *You need to spend a large percentage (at least 20%) of your prep time on IBC Chapter 16 and become familiar with it. You do not need to force yourself to memorize the numbers and all the detail. Just reading it a few times and becoming familiar with the information should be adequate.*

2. *Architectural Graphic Standards*
   Ramsey, Charles George, and John Ray Hoke Jr. *Architectural Graphic Standards.* The American Institute of Architects & Wiley, latest edition. You may have a few questions asking you what some basic graphic symbols mean. This is a good book to skim through.

3. **FREE information on truss and beam diagrams** can be found at the following link:
   http://ocw.mit.edu/ans7870/4/4.463/f04/module/Start.html

4. **The FREE PDF file of FEMA publication number 454 (FEMA454),** *Designing for Earthquakes: A Manual for Architects*, is available at the following link:
   http://www.fema.gov/library/viewRecord.do?id=2418

Note:
*You need to spend a large percentage (at least 30%) of your prep time on this **FEMA454** PDF file. Focus on **Chapters 4, 5, 8, and 9**. You may have **many real ARE questions** based on these chapters. This PDF book has many diagrams and photos and helps you understand what happened in buildings that failed during an earthquake.*

5. **The FREE PDF file of *Wind Design Made Simple*** by ICC TRI-Chapter Uniform Code Committee is available at the following link:
   http://www.calbo.org/Documents/SimplifiedWindHandout.pdf

   Note:
   You need to become familiar with this file. You do not need to force yourself to memorize the numbers and all the details, just reading it a few times and becoming familiar with the information should be adequate.

6. Arnold, Christopher. ***Building at Risk***, is available for FREE at the AIA website:
   http://www.aia.org/aiaucmp/groups/aia/documents/pdf/aiap016810.pdf

7. **Construction Specifications Institute (CSI) MasterFormat & *Building Construction***
   Become familiar with the new 6-digit CSI Construction Specifications Institute (CSI) MasterFormat as there may be a few questions based on this publication. Make sure you know which items/materials belong to which CSI MasterFormat specification section, and memorize the major section names and related numbers. For example, Division 9 is Finishes, and Division 5 is Metal, etc. Another one of my books, *Building Construction*, has detailed discussions on CSI MasterFormat specification sections.

Mnemonics for the 2004 CSI MasterFormat

The following is a good mnemonic, which relates to the 2004 CSI MasterFormat division names. Bold font signals the gaps in the numbering sequence.

This tool can save you lots of time: if you can remember the four sentences below, you can easily memorize the order of the 2004 CSI MasterFormat divisions. The number sequencing is a bit more difficult, but can be mastered if you remember the five bold words and numbers that are not sequential. Memorizing this material will not only help you in several divisions of the ARE, but also in real architectural practice

Mnemonics (pay attention to the underlined letters):
Good students can memorize material when teachers order.
F students earn F's simply 'cause **forgetting** principles have **an** effect. (21 and 25)
C students **end** everyday understanding things without memorizing. (31)
Please make professional pollution prevention inventions **everyday**. (40 and 48)

1-Good.................................... General Requirements
2-Students............................. (Site) now Existing Conditions
3-Can......................................Concrete
4-Memorize...........................Masonry
5-Material ....Metals
6-When...................................Woods and Plastics
7-Teachers.............................Thermal and Moisture
8-Order..................................Openings

9-F..Finishes
10-Students............................ Specialties
11-Earn..................................Equipment
12-F's.....................................Furnishings
13-Simply..............................Special Construction
14-'Cause...............................Conveying
21-Forgetting .... **Fire**
22-Principles.........................Plumbing
23-Have................................ HVAC
25-An......................................**Automation**
26-Effect............................... Electric

27-C....................................... Communication
28-Students............................ Safety & Security
31-End.....................................**Earthwork**
32-Everyday..........................Exterior
33-Understanding ....Utilities
34-Things.............................. Transportation
35-Without Memorizing........ Waterways and Marine

40-Please..............................**Process Integration**
41-Make................................ Material Processing and Handling Equipment
42-Professional.....................Process Heating, Cooling, and Drying Equipment
43-Pollution.......................... Process Gas and Liquid Handling, Purification and Storage Equipment
44-Prevention........................Pollution Control Equipment
45-Inventions........................Industry-Specific Manufacturing Equipment
48-Everyday.........................**Electrical Power Generation**

Note:

There are 49 CSI divisions. The "missing" divisions are those "reserved for future expansion" by CSI. They are intentionally omitted from the list.

C. Overall Strategies, Tips, and Solutions to the Official NCARB SS Practice Program Problem

1. Overall strategies

To most candidates, the Multiple Choice (MC) portion of an ARE division is harder than the graphic vignette portion. Some of the MC questions are based on experience and you do NOT have a set of fixed study materials for them. You WILL make some mistakes on the MC questions no matter how hard you study.

On the other hand, the graphic vignette is relatively easier, and there are good ways to prepare for this. You should really take the time to study and practice the NCARB graphic software well. Try to **nail the graphic vignette** perfectly. This way, you will have a better chance of passing even if you answer some MC questions incorrectly.

Tips: *Most people do poorly on the MC portion of the SS division, especially those who do NOT have a lot of working experience, but curiously not too many people fail because of the MC portion. Most people fail the ENTIRE SS section because they have made **one** fatal mistake on the graphic vignette section. So, practice the NCARB SS practice program graphic software and make sure you absolute NAIL the vignette section. This is key for you to pass.*

The official NCARB SS exam guide gives a passing and failing solution to the sample vignette, but it does NOT show you the step-by-step details.

We are going to fill in the blanks here and offer you step-by-step instructions, command by command.

You really need to spend time practicing to become VERY familiar with NCARB's graphic software. This is because all ARE graphic vignettes are timed, and you do NOT have the luxury to think about how to use the software during the exam. Otherwise, you may NOT be able to finish your solution in time.

The following solution is based on the official NCARB SS practice program for the **ARE 4.0**. Future versions of ARE may have some minor changes, but the principles and fundamental elements should be the same. The official NCARB SS practice program has not changed much since its introduction and the earlier versions are VERY similar to, if not exactly the same as, the current ARE 4.0. The actual graphic vignette of the SS division should be VERY, VERY similar to the practice one on NCARB's website.

2. Tips

1) You need to install the NCARB SS practice program, and become familiar with it. I am NOT going to repeat the vignette description and requirements here since they are already written in the NCARB practice program.

   See the following link for a FREE download of the NCARB practice program:
   http://www.ncarb.org/ARE/Preparing-for-the-ARE.aspx

2) Review the general test directions, vignette directions, program, and tips carefully.
3) Press the space bar to go to the work screen.
4) Read the program and codes in the NCARB Exam Guide several times the week before your exam. Become VERY familiar with this material, and you will be able to read the problem requirements MUCH faster during the real exam because you can immediately identify which criteria are different from the practice exam.

3. Step-by-step solution to the official NCARB SS practice program problem: graphic vignette section

1) Let us start with the lower level. Click on **layers** to make sure we set the current layer to the lower level (figure 2.1).

   Note: Once you draw an element on a level, it is impossible to move it to another level. So, this step is VERY important. If you draw elements on the wrong level, you may waste a lot of time and not have enough time to redraw them on the correct level, causing you to fail the vignette and the SS ARE division.

2) Click on **cursor** to set the cursor to full-screen mode (figure 2.2). This will make it easier for you to align structural elements.

3) Use **Draw > Bearing Wall w/ Bond Beam** to draw the bearing walls (figure 2.3). Make sure you do not accidentally cover the openings.

4) Use **Draw > Beam or Lintel** to place lintels over all openings in the bearing walls (figure 2.4). Make sure that each lintel is supported by bearing walls on both ends.

5) The window wall and the clerestory window extend to the underside of the structure above. All other openings (including the opening between the Common Area and Covered Entry) have a head height of 7 ft above finished floor. Therefore, we need to add a beam or lintel for the opening between the Common Area and Covered Entry. Use **Draw > Beam or Lintel** to place a long lintel over this opening and span it between two bearing walls (figure 2.5). This beam will also support the upper wall between the Common Area and Covered Entry.

   Note: You can also make the wall between the Common Area and Covered Entry a bearing wall, and use a shorter lintel.

6) Use **Draw > Joists > Select joist direction > Spacing at 48" o.c.** to draw joists between the bearing walls (figure 2.6).

Note:
- *The joists are drawn as a 2-point rectangle.*
- *Make sure both ends of the joists are supported by bearing walls, beams, or lintels.*
- *Make sure the joists support the edge of the roof at locations with no bearing walls, beams, or lintels.*
- *Per the program, the metal roof deck is capable of carrying the design loads on spans up to and including 4 ft. Therefore, the maximum spacing between the joists is 4 ft or 48". We set the spacing at 48".*

7) Use **Draw > Decking > Select deck direction** to draw decks between the joists (figure 2.7).

Note:
- *The decks are drawn as a 2-point rectangle.*
- *Make sure that both ends of the decks are supported by joists.*
- *The grey-color double arrow indicates the direction of the deck. It should be perpendicular to the joists.*

8) The lower level is complete. Let us start with the upper level. Click on **layers** to open a dialogue box, set the **Current Level** to **Upper,** and then click **OK** (figure 2.8).

9) Use **Draw > Column** to draw six columns on top of the bearing walls. Align the columns (figure 2.9).

Note:
- *Once you set the **Current Level** to **Upper** and select the other level as visible, the elements on the lower level show in grey color.*
- *The column layout accommodates the clerestory window located along the full length of the north wall of the common area.*

10) Use **Draw > Beam or Lintel** to place beams between the columns along the direction of the bearing walls (figure 2.10).

Note:
Place beams from the center of a column to the center of an adjacent column.

11) Use **Draw > Joists > Select joist direction > Spacing at 48" o.c.** to draw joists between the bearing beams (figure 2.11).

12) Use **Draw > Decking > Select deck direction** to draw decks between the joists (figure 2.12).

13) This is your final solution (figure 2.13).

14) Of course, you can also solve this vignette using only columns and no bearing walls. Just make sure your upper level columns are supported by lower level columns at the same location (figure 2.14 & figure 2.15).

Note:
Pay attention to the beam on the east wall of the Common Area at the lower level. We need it to support the wall above.

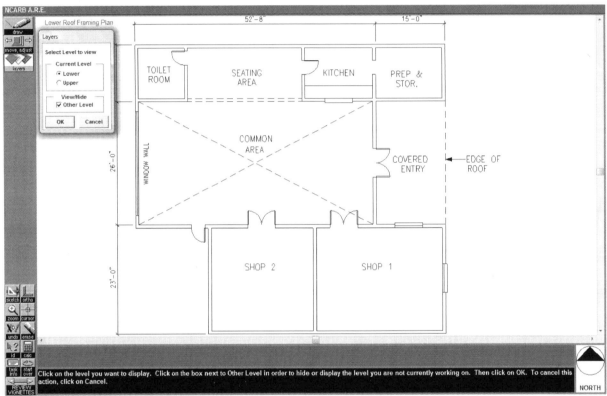

Figure 2.1   Click on **layers** to make sure we set the current layer to the lower level.

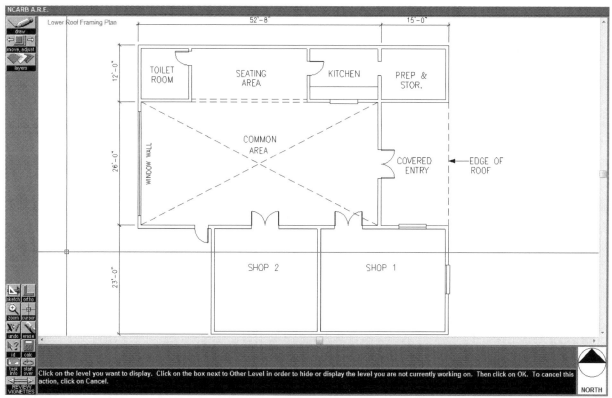

Figure 2.2   Click on **cursor** to set the cursor to full-screen mode.

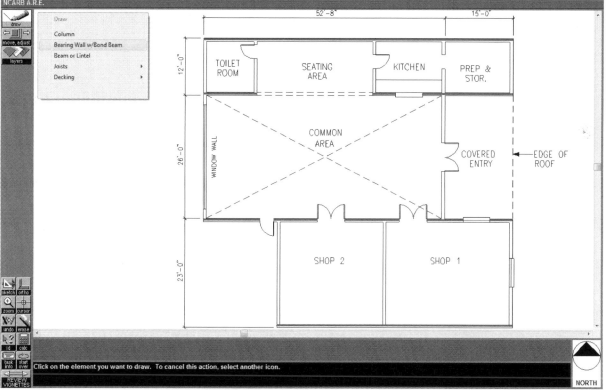

Figure 2.3   Use **Draw > Bearing Wall w/ Bond Beam** to draw the bearing walls.

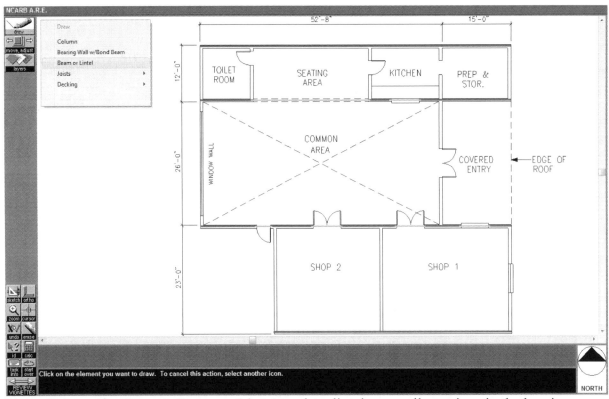

Figure 2.4   Use **Draw > Beam or Lintel** to place lintels over all openings in the bearing walls.

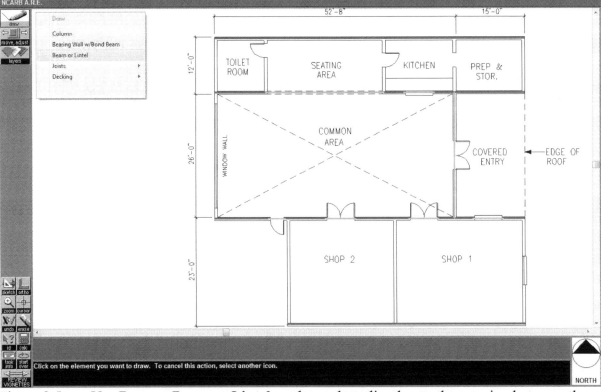

Figure 2.5   Use **Draw > Beam or Lintel** to place a long lintel over the opening between the Common Area and Covered Entry and span it between two bearing walls.

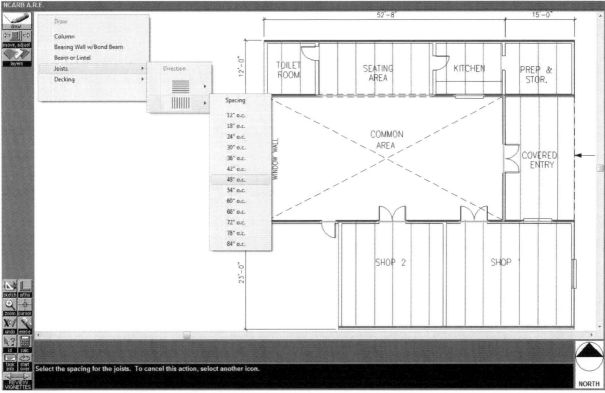

Figure 2.6   Use **Draw > Joists > Select joist direction > Spacing at 48" o.c.** to draw joists between the bearing walls.

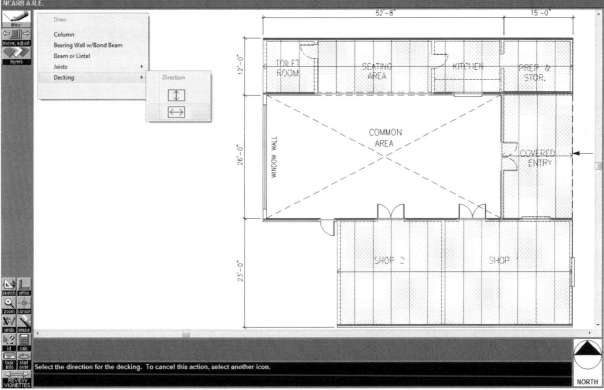

Figure 2.7   Use **Draw > Decking > Select deck direction** to draw decks between the joists.

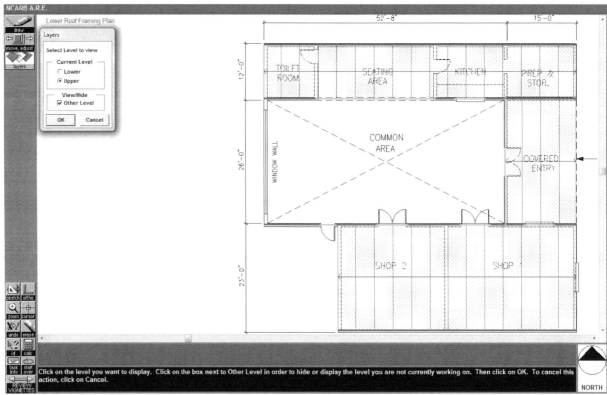

Figure 2.8   Click on **layers** to open a dialogue box, set the **Current Level** to **Upper,** and then click **OK.**

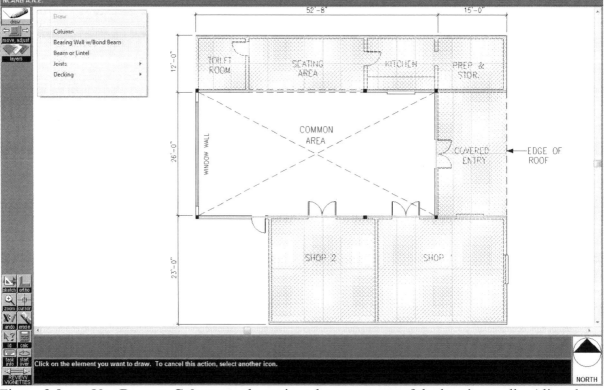

Figure 2.9   Use **Draw > Column** to draw six columns on top of the bearing walls. Align the columns.

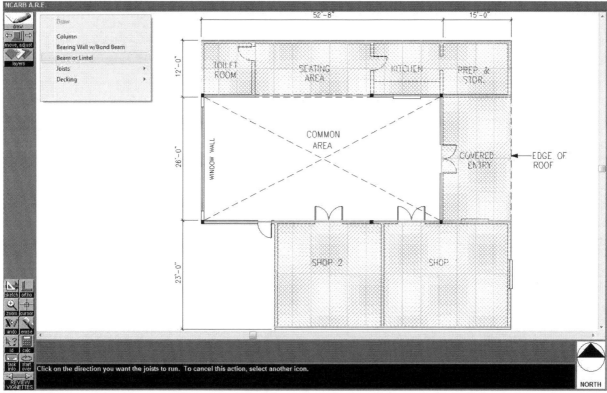

Figure 2.10   Use **Draw > Beam or Lintel** to place beams between the columns along the direction of the bearing walls.

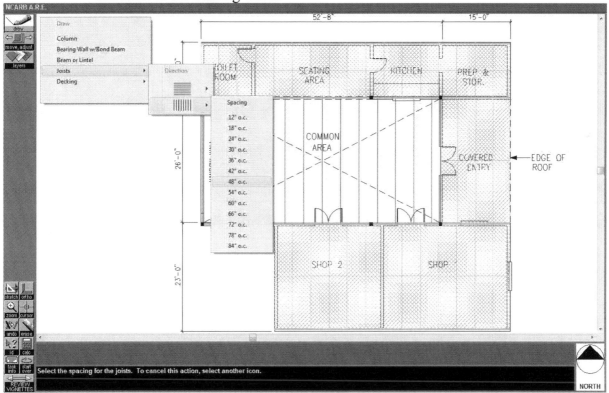

Figure 2.11   Use **Draw > Joists > Select joist direction > Spacing at 48" o.c.** to draw joists between the bearing beams.

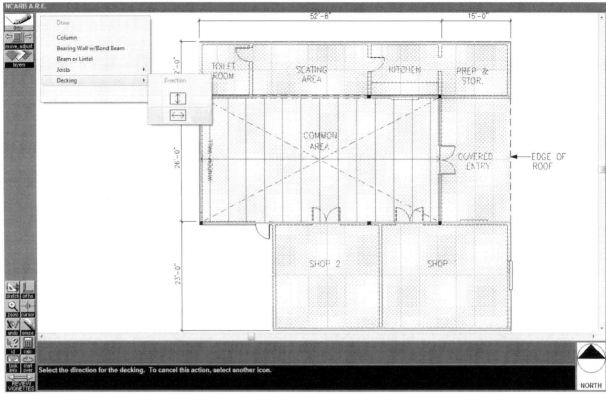

Figure 2.12   Use **Draw > Decking > Select deck direction** to draw decks between the joists.

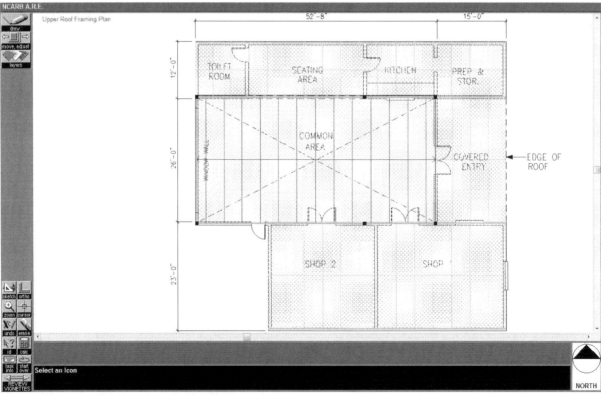

Figure 2.13   This is your final solution.

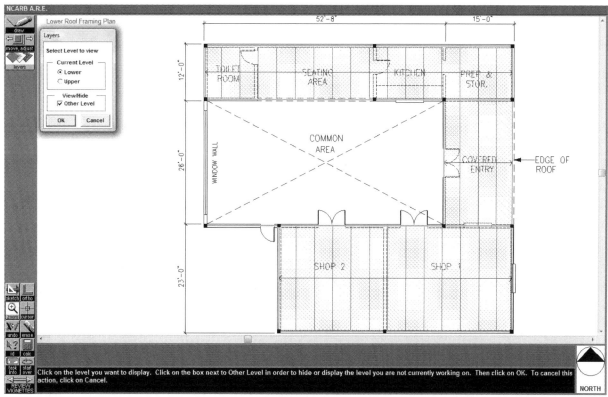

Figure 2.14   Use only columns and no bearing walls to solve the vignette: lower level.

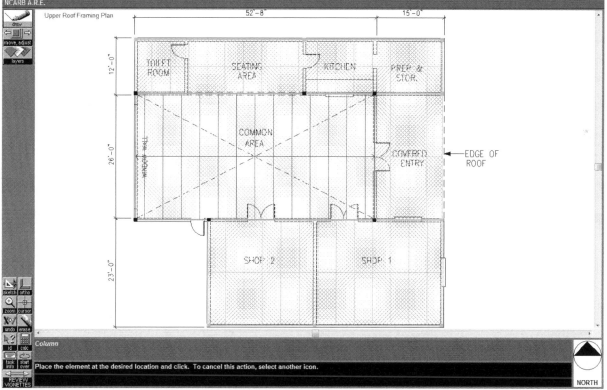

Figure 2.15   Use only columns and no bearing walls to solve the vignette: upper level.

Chapter Three

ARE Mock Exam for
Structural Systems (SS) Division

A. Mock Exam: SS Multiple-Choice (MC) Section

1. Per the ***International Building Code (IBC)***, which of the following is correct? **Check the two that apply.**
   a. *IBC* allows live load reduction in most cases.
   b. *IBC* allows live load reduction in a few cases.
   c. *IBC* does not allow live load reduction for public assembly occupancy with a live load equal to or less than 100 psf.
   d. *IBC* does not allow live load reduction for a live load equal to or more than 100 psf.

2. The load of an automobile moving in a parking garage is a
   a. dynamic load
   b. impact load
   c. dead load
   d. vertical load

3. Which of the following is incorrect? **Check the two that apply.**
   a. Wind velocity affects wind loading.
   b. Wind velocity does not affect wind loading.
   c. Wind velocity is higher on the 10th floor than on the ground floor.
   d. Wind velocity is lower on the 10th floor than on the ground floor

4. If a load of 10,000 lbf is applied to a 10' long, 4x4 Douglas fir no. 2 wood, it will compress _____ inch. The modulus of elasticity for Douglas fir no. 2 is 1,700,000 psi.

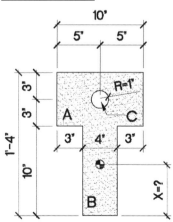

Figure 3.1   Locate the centroid of the hatched area

5. Locate the centroid of the hatched area as shown in figure 3.1 and calculate to find that x = _____ .

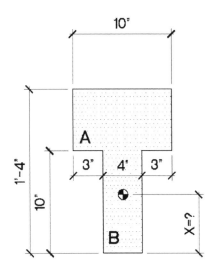

Figure 3.2   The moment of inertia for a composite section

6. The moment of inertia for the composite section as shown in figure 3.2 is _____ ft-lbf.

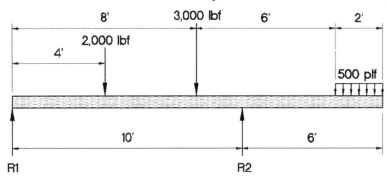

Figure 3.3   The reactions of a beam

7. The reactions of the beam shown in figure 3.3 are
    a. R_1 = 1,200 lbf, R_2 = 4,600 lbf
    b. R_1 = 4,600 lbf, R_2 = 1,200 lbf
    c. R_1 = 1,300 lbf, R_2 = 4,700 lbf
    d. R_1 = 4,700 lbf, R_2 = 1,300 lbf

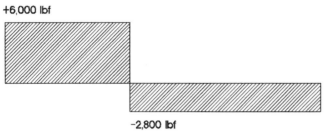

+6,000 lbf

−2,800 lbf

Figure 3.4   What is this diagram?

8. Figure 3.4 shows
   a. a shear diagram of uniformly distributed loads
   b. a shear diagram of concentrated loads
   c. a load diagram
   d. none of the above

9. The moment diagram of a uniformly distributed load is
   a. a rectangle
   b. a triangle
   c. a trapezoid
   d. a parabolic shape

10. The location where the line of a moment diagram crosses zero is the
    a. inflection point
    b. deflection point
    c. zero point
    d. balance point

11. Which of the following is a method used to determine the forces in truss members? **Check the three that apply.**
    a. method of analysis
    b. method of joints
    c. method of sections
    d. method of moment
    e. graphic method

12. A standard penetration test (SPT) tests the
    a. ability of water to penetrate soil
    b. ability of water to penetrate retaining walls
    c. density and consistency of soil
    d. density and consistency of slab

13. USCS stands for
    a. Unified Standard Classification System
    b. Unified Soils Classification System
    c. Unified States Concrete Society
    d. none of the above

14. There are three types of retaining walls. **Check the three that apply.**
    a. gravity wall
    b. brace wall
    c. cantilever wall
    d. counterfort wall
    e. CMU wall

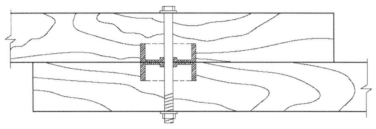

Figure 3.5   What does this diagram show?

15. Figure 3.5 shows a
    a. split ring connector
    b. shear plate connector
    c. double shear connector
    d. multiple shear connector

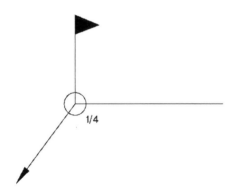

1/4

Figure 3.6   What does this diagram show?

16. Figure 3.6 shows a
    a. field weld
    b. flare weld
    c. bevel weld
    d. square weld

17. The concerns for a roof without sufficient slope include (**Check the two that apply.**)
    a. ponding
    b. deflection
    c. inflection
    d. failure in an earthquake

18. Which of the following cannot be directly embedded in a concrete foundation or slab?
    a. ABS pipe
    b. PVC pipe
    c. aluminum conduit
    d. copper conduit

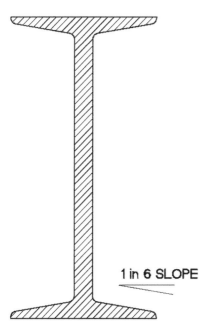

1 in 6 SLOPE

Figure 3.7   What does this diagram show?

19. The structural steel shown in figure 3.7 is a
    a. "C" shape
    b. "ST" shape
    c. "S" shape
    d. "W" shape

20. Which of the following steel columns will fail first by buckling under loads?
    a. a steel column with a slenderness ratio of 32
    b. a steel column with a slenderness ratio of 28
    c. a steel column with a slenderness ratio of 26
    d. a steel column with a slenderness ratio of 18

21. If end conditions are the only differences, which of the following columns has the highest slenderness ratio?
    a. a steel column fixed in both rotation and translation at both ends
    b. a steel column fixed in both rotation and translation at one end, but free to rotate and fixed in translation at the other

    c. a steel column fixed in both rotation and translation at one end, but free to translate and fixed in rotation at the other

    d. a steel column fixed in both rotation and translation at one end, but free to rotate and translate at the other

    e. a steel column free to rotate but fixed in translation at both ends

22. A 16-foot-high W12 x 120 column is fixed in both rotation and translation at both ends. The radius of gyration is 5.51 inches in the X-X-axis and 3.13 inches the Y-Y-axis. K = 0.65. What is the slenderness ratio?

    a. 38.87

    b. 39.87

    c. 40.87

    d. 41.87

23. UFER is an item for

    a. civil work

    b. structural work

    c. electrical work

    d. mechanical work

    e. plumbing work

24. A contractor is working on a single-family home project. Which subcontractor must complete the underground work before s/he can pour the concrete slab and foundation? **Check the three that apply.**

    a. framing subcontractor

    b. electrical subcontractor

    c. mechanical subcontractor

    d. plumbing subcontractor

    e. plaster subcontractor

25. Open-web steel joists spaced 4 ft on center span 36 ft. The dead load = 30 psf (including the weight of the joists), live load = 50 psf, and maximum allowable deflection = 1/360 of the span. Based on table 3.1, what is the most economical section to use?

Table 3.1   Partial Load Table for Open-Web Steel Joists, K-Series
(Based on a maximum allowable tensile stress of 50,000 psi;
loads in pounds per linear foot)

Joist Designation	20K9	20K10	22K7	22K9
Depth (in)	20	20	22	22
Approx. Wt. (lbs/ft)	10.8	12.2	9.7	11.3
Span (ft) ↓				
35	329	390	303	364
	179	210	185	219
36	311	369	286	344
	164	193	169	201

    a. 20K9
    b. 20K10
    c. 22K7
    d. 22K9

26. What is the total horizontal load exerted on a 4-foot-high retaining wall? The wall is retaining inorganic silts and clayey silts with an active pressure of 45 psf/ft.
    a. 240 psf
    b. 360 psf
    c. 480 psf
    d. 600 psf

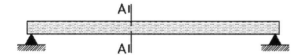

Figure 3.8   A simply supported beam

27. The section A-A of a simply supported beam shown in figure 3.8 has (**Check the two that apply.**)
    a. tension stresses in the top half of the beam
    b. compression stresses in the top half of the beam
    c. tension stresses in the bottom half of the beam
    d. compression stresses in the bottom half of the beam

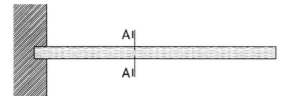

Figure 3.9   A cantilever beam

28. Section A-A of the cantilever beam shown in figure 3.9 has (**Check the two that apply.**)
    a. tension stresses in the top half of the beam
    b. compression stresses in the top half of the beam
    c. tension stresses in the bottom half of the beam
    d. compression stresses in the bottom half of the beam

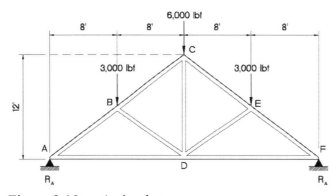

Figure 3.10   A simple truss

29. Figure 3.10 shows a simple truss. Using the method of joints, the force in member AB (the member between joints A and B) is
    a. 10,000 lbf
    b. 11,000 lbf
    c. 12,000 lbf
    d. 13,000 lbf

30. For the same simple truss in figure 3.10, using the method of sections, the force in member BC is
    a. 6,500 lbf
    b. 7,500 lbf
    c. 8,500 lbf
    d. 9,500 lbf

31. A contractor is working on a wood frame residential project with a slab on grade and continuous footing. Which of the following statements is correct? **Check the two that apply.**
    a. The hold-downs bolts should be placed before pouring the concrete footing and slab.
    b. The hold-downs bolts should be placed after pouring the concrete footing and slab.
    c. The anchor bolts should be placed before pouring the concrete footing and slab.

d. The anchor bolts should be placed after pouring the concrete footing and slab.

32. An architect is working on a residential project. The structural plans call for concrete with a compressive strength of 2,500 psi. The plan checker requests the use of concrete with a compressive strength of 4,500 psi because the potential for sulfate at the project site. What should the architect do?
   a. Ask the structural engineer to change the plans to use 4,500 psi concrete.
   b. Immediately respond to the plan check correction and refuse to use the 4,500 psi concrete because it is not required by the structural design.
   c. Ask the structural engineer to change the plans to use 4,500 psi concrete, but note on the structural plans "4,500 psi concrete is not required by structural design, special inspection is not required."
   d. Ask the structural engineer to change the plans to use 4,500 psi concrete, and note on the structural plans "special inspection is required."

33. Which of the following structural members is typically Douglas fir #1 instead of #2? **Check the two that apply.**
   a. 2x10 floor joists
   b. 4x6 columns
   c. 4x6 window and door headers
   d. 4x8 beams

Figure 3.11   A structural item

34. Figure 3.11 shows a (an)
   a. lag screw
   b. anchor bolt
   c. hold-down bolt
   d. base plate screw

Figure 3.12   A structural item

35. Figure 3.12 shows a (an)
    a. lag screw
    b. anchor bolt
    c. hold-down bolt
    d. base plate screw

36. Number 6 rebar is _____ inch in diameter.

37. _____ are used to reduce the amount of water needed for mixing concrete.
    a. Accelerators
    b. Air-entraining agents
    c. Plasticizers
    d. Fine aggregates

Figure 3.13   A structural item

38. Figure 3.13 shows a
    a. column base
    b. strap
    c. hold-down
    d. structural plate

39. The diameter of #3 rebar is
    a. 3/32"
    b. 3/16"
    c. 3/8"
    d. 3/4"

40. Which of the following are two different designations for the same welded wire fabric? **Select the two that apply.**
    a. 6x6-10/10
    b. 6x6-W1.0x1.0
    c. 6x6-W1.2x1.2
    d. 6x6-W1.4x1.4
    e. 6x6-W1.6x1.6

41. Which of the following diagrams in figure 3.14 show correct placement of rebar for the concrete beams of a gas station canopy supported by concrete columns? (Some elements are not shown for clarity.)
    a. A
    b. B
    c. C
    d. D

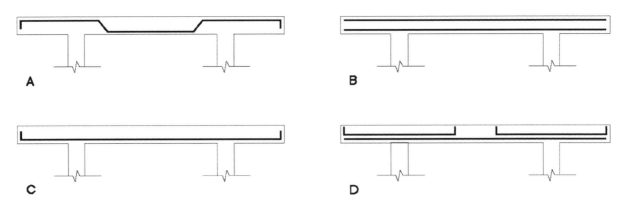

Figure 3.14   Rebar for beams at a concrete canopy

42. The gradient height for metropolitan areas is approximately
    a. 900 ft
    b. 1200 ft
    c. 1500 ft
    d. 1800 ft

43. Using Method II, which of the following diagrams in figure 3.15 show the correct direction of wind pressure?
    a. A
    b. B
    c. C
    d. D

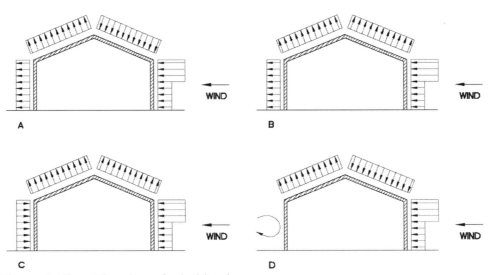

Figure 3.15   Direction of wind loads

44. Which of the following diaphragms in figure 3.16 is a correct diagram of wind loads?
    a. A
    b. B
    c. C
    d. D

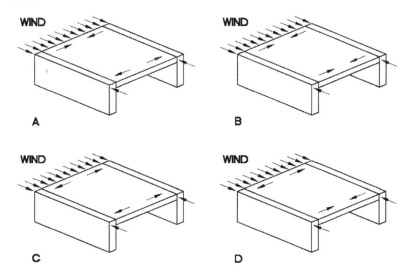

Figure 3.16   Wind loads acting on a diaphragm

45. In reference to figure 3.17, the brace at location A is in
    a. tension
    b. compression
    c. tension on top and compression at the bottom
    d. none of the above

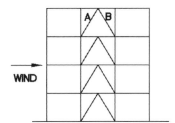

Figure 3.17   K-brace

46. Site Class A, B, C, D, E, or F in Chapter 16 of IBC is based on
    a. earthquake zones
    b. wind load zones
    c. soil properties
    d. none of the above

47. For a hospital, the Importance Factor (I Factor) for snow is:
    a. 1.15
    b. 1.2
    c. 1.25
    d. 1.5

48. The purpose of camber in a beam is
    a. to compensate for deflection
    b. to compensate for torsion
    c. to compensate for tension
    d. to compensate for compression

49. A soft story is most likely to occur in a building at
    a. the first floor
    b. the second floor
    c. the middle floor
    d. the top floor

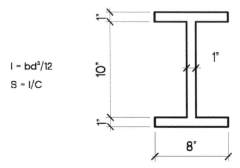

Figure 3.18   Section modulus

50. The section modulus for the geometric section as shown in figure 3.18 is_____
    _____in³.

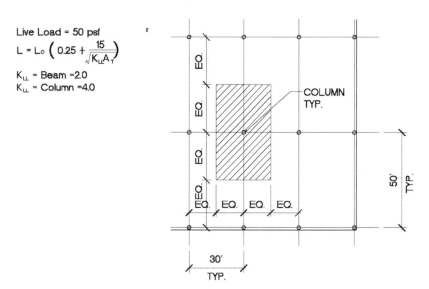

Figure 3.19   The live load for the floor supported by a column

51. The live load for the floor supported by the column as shown in figure 3.19 is
    _____kips (kg).

52. The tendency of a solid material to move slowly or deform permanently under the influence of sustained stress is called
    a. shrinkage
    b. temperature expansion
    c. creep
    d. contraction

53. When the ratio of a slab's length (long direction to short direction) is greater than _____ it can be considered a one-way slab.
    a. 3:2
    b. 2:1
    c. 5:2
    d. 3:1

54. A 9-story hotel has an elevator. Which of the following is the minimum increase in live load to accommodate for elevator impact according to IBC ?
    a. 0 percent
    b. 33 percent
    c. 50 percent
    d. 100 percent

55. Which of the following in figure 3.20 is an eccentrically braced frame (EBF)?

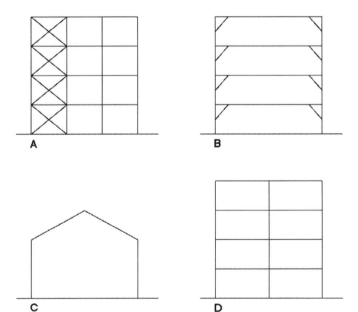

Figure 3.20   Pick an eccentrically braced frame (EBF)

56. For the flexible diaphragm shown in figure 3.21, what is the shear at F_2?
    a. 15 kips
    b. 20 kips
    c. 25 kips
    d. 30 kips

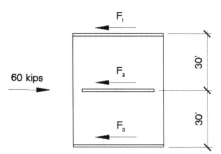

Figure 3.21   A flexible diaphragm

57. If all floors have the same area and similar layout, base isolation is most effective for which of the following building heights?
    a. one-story
    b. six-story
    c. thirty-story
    d. sixty-story

58. Which of the following affects the slenderness ratio of a column? **Check the three that apply.**
    a. end-constraints
    b. size of the member
    c. bracing
    d. radius of gyration
    e. rotating the column

59. If an apartment building fails in an earthquake, who of the following is likely to have primary legal responsibilities? **Check the three that apply.**
    a. the owner
    b. the apartment manager
    c. the building official
    d. the contractor
    e. the architect
    f. the structural engineer
    g. the soils engineer

60. If resisted only by gravity forces, what is the factor of safety against overturning for a concrete shear wall as shown in figure 3.22? Suppose the weight of concrete equals 150 lb/ft³ [23.5 kN/m3], and the dead load equals 300 kips [1325 kN].
Disregard the weight of the soil over the footing.
   a. 5.91
   b. 6.91
   c. 7.91
   d. 8.91

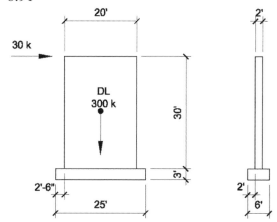

Figure 3.22   The factor of safety against overturning for a concrete shear wall

61. Which of the following connections has the least structural value? **Check the two that apply.**
   a. bond beams
   b. high-strength mortar
   c. metal plate strap anchors
   d. running bond
   e. stacked bond
   f. steel dowels

62. The abrupt changes in ground shear distribution at the base of a sufficiently tall structure subject to critical lateral displacement is called
   a. moment shift
   b. shear interruption
   c. shear distribution
   d. P-Delta effect
   e. overturning moment

63. Per a soils report, the soil bearing capacity is 3,000 psf [143 500 N/m2]. If the applied load is 60,000 lbs [265 kN] for a column, and the pad footing is square, what is the width of the pad footing?
   a. 3'-0"
   b. 3'-6"
   c. 4'-0"
   d. 4'-6"

64. The minimum concrete coverage for rebar in a spread footing is
    a. 1"
    b. 1 ½"
    c. 2"
    d. 3"

65. Which of the following is NOT an effective structural system to resist lateral loads?
    a. shear wall
    b. frame tube
    c. hinged frame
    d. portal frame
    e. moment-resisting frame

66. Which of the following is a method used to determine wind loads on a structure? **Check the two that apply.**
    a. normal force method
    b. lateral force method
    c. projected area method
    d. diagram method

67. In regard to wind load design, the categories B, C, and D are based on
    a. surface roughness
    b. geography
    c. distance to the ground surface
    d. wind strength

68. Which of the following in figure 3.23 is the correct diagram of wind pressure for a 12-foot high building with a 30-foot square plan?
    a. A
    b. B
    c. C
    d. D

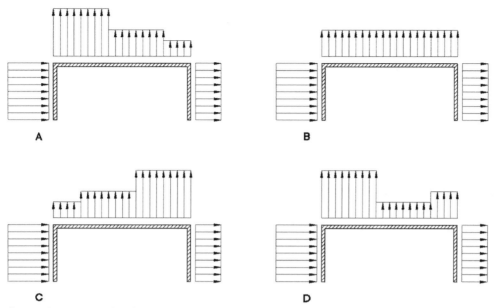

Figure 3.23   Selecting a correct diagram of wind pressure

69. With regard to the 6-story building in figure 3.24, which location has the greatest shear during an earthquake?
    a. 1
    b. 3
    c. 6
    d. 7

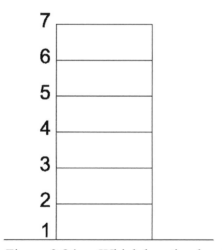

Figure 3.24   Which location has the greatest shear during an earthquake?

70. Which of the following is correct? **Check the two that apply.**
    a. The Modified Mercalli Scale is more accurate than the Richter Scale.
    b. The Richter Scale is more accurate than the Modified Mercalli Scale.
    c. In the Richter Scale, an earthquake of magnitude 8 releases about 10 times more energy than one of magnitude 7.
    d. In the Richter Scale, an earthquake of magnitude 8 releases about 32 times more energy than one of magnitude 7.
    e. In the Modified Mercalli Scale, an earthquake of magnitude 8 releases about 10 times more energy than one of magnitude 7.
    f. In the Modified Mercalli Scale, an earthquake of magnitude 8 releases about 32 times more energy than one of magnitude 7.

71. The Seismic Design Category (SDC) of a structure is determined by its (**Check the three that apply.**)
    a. design spectral response acceleration coefficients
    b. occupancy category
    c. height
    d. weight
    e. footprint

72. Which of the following regarding the Seismic Design Category (SDC) of a structure is true? **Check the two that apply.**
    a. Buildings in SDC A need to meet more restrictive earthquake requirements than buildings in SDC C.
    b. Buildings in SDC C need to meet more restrictive earthquake requirements than buildings in SDC A.
    c. SDC includes categories A through D.
    d. SDC includes categories A through E.

73. Which of the following regarding the earthquake's impact on a structure is NOT true? **Check the two that apply.**
    a. In general, a number of cycles of moderate acceleration, sustained over time, can be much more difficult for a building to withstand than a single much larger peak.
    b. In general, a number of cycles of moderate acceleration, sustained over time, are much easier for a building to withstand than a single much larger peak.
    c. Low–frequency waves (higher than 10 hertz) tend to have high amplitudes of acceleration but small amplitudes of displacement, compared to high-frequency waves, which have small accelerations and relatively large velocities and displacements.
    d. High–frequency waves (higher than 10 hertz) tend to have high amplitudes of acceleration but small amplitudes of displacement, compared to low-frequency waves, which have small accelerations and relatively large velocities and displacements.

74. The **bracketed duration** is defined as the time between the first and last peak of motion that exceeds a threshold value, commonly taken as
   a. 0.03g
   b. 0.05g
   c. 0.3g
   d. 0.5g

75. Which of the following is the most important determinant of a building's period?
   a. construction materials
   b. contents
   c. geometric proportions
   d. height
   e. structural system
   f. weight

76. The natural period of the ground varies from about _____, depending on the nature of the ground.
   a. 0.1 seconds to 0.3 seconds
   b. 0.4 seconds to 2 seconds
   c. 2 seconds to 3 seconds
   d. 3 seconds to 4 seconds

77. When a vibrating or swinging object is given further pushes that are also at its natural period, its vibrations increase dramatically in response to even rather small pushes. This phenomenon is called
   a. acceleration
   b. inertia
   c. resonance
   d. velocity increase

78. Which of the following statements regarding the earthquake's impact on a structure is true? **Check the two that apply.**
   a. Accelerations created by ground motion increase rapidly as the building damping value decreases.
   b. Accelerations created by ground motion decrease rapidly as the building damping value decreases.
   c. Damping is measured by reference to a theoretical acceleration level termed critical acceleration.
   d. Damping is measured by reference to a theoretical damping level termed critical damping.

79. Which of the following statements regarding earthquakes is true? **Check the two that apply.**
    a. P wave is a wave that moves perpendicular to the ground surface.
    b. P wave is a primary wave that alternately pushes (compresses) and pulls (dilates) the rock.
    c. P waves, just like acoustic waves, are able to travel through solid rock, such as granite and alluvium; through soils; and through liquids, such as volcanic magma or the water of lakes and oceans.
    d. S waves, just like acoustic waves, are able to travel through solid rock, such as granite and alluvium; through soils; and through liquids, (such as volcanic magma or the water of lakes and oceans.

80. Which of the following terms is used to define the earthquake wave? **Check the two that apply.**
    a. H wave
    b. Love wave
    c. Rayleigh wave
    d. W wave

81. At a field visit, an architect notices that the contractor has completely filled the space between two structural columns with a rigid wall per the owner's instruction. What can the architect do? **Check the two that apply.**
    a. Advise the owner to have the structural engineer review the changes and submit revised plans to the city.
    b. Do nothing because this is a very minor revision, and it has no impact on the structural integrity of the building, and does not even require a building permit.
    c. Advise the owner that this may have created a structural problem and order the contractor to rip out the wall after the client agrees this is an unnecessary element.
    d. Revise the architectural plans and submit the architectural plans to the city since the revision does not involve structural plans.

82. Which of the following is the governing factor in the ultimate seismic behavior of a particular structure?
    a. the weight of the structure
    b. the height of the structure
    c. the configuration of the structure
    d. none of the above

83. Which of the following is the correct order to arrange structural systems from low post-earthquake repair cost to high post-earthquake repair cost?
    a. seismic isolation, dampers plus steel moment-resisting frame, un-bonded steel brace, timber framing, ductile steel moment-resisting frame, steel frame plus braces
    b. seismic isolation, dampers plus steel moment-resisting frame, steel frame plus braces, un-bonded steel brace, timber framing, ductile steel moment-resisting frame
    c. seismic isolation, dampers plus steel moment-resisting frame, steel frame plus braces, ductile steel moment-resisting frame, un-bonded steel brace, timber framing
    d. seismic isolation, dampers plus steel moment-resisting frame, ductile steel moment-resisting frame, un-bonded steel brace, timber framing, steel frame plus braces

84. The three basic alternative types of vertical lateral force–resisting systems are _____ _____.

85. The term used to identify a horizontal-resistance member that transfers lateral forces between vertical-resistance elements (shear walls or frames) is
    a. beam
    b. space frame
    c. diaphragm
    d. floor system

86. Which of the following contribute primarily to poor seismic performance and occasional failure?
    a. regular building shapes
    b. architectural irregularities
    c. weights of the buildings
    d. heights of the buildings

87. Which of the following contribute to poor seismic performance and occasional failure?
    a. re-entrant corners
    b. soft or weak floors
    c. torsion
    d. short-column phenomenon
    e. all of the above

88. Which of the following do NOT have the potential to seriously impact seismic performance?
    a. soft and weak stories
    b. discontinuous shear walls
    c. variations in perimeter strength and stiffness
    d. re-entrant corners
    e. circular plans

89. The International Style often has a number of characteristics not present in earlier frame and masonry buildings that has led to poor seismic performance, including: (**Check the three that apply.**)
    a. the elevation of the building on stilts or pilotis
    b. the free plan and elimination of interior-load bearing walls
    c. the great increase of exterior glazing and the invention of the light-weight curtain wall
    d. the building plan configuration
    e. the elimination of building decorations and related redundancy of building components

90. When was the first seismic code created in the United States?
    a. 1900s
    b. 1910s
    c. 1920s
    d. 1930s

91. Technical performance levels translate qualitative performance levels into damage states expected for structural and nonstructural systems. Table 6-2 of the Structural Engineers Association of California (SEAOC) Vision 2000 document proposes four qualitative performance levels. **Check the four that apply.**
    a. fully operational
    b. operational
    c. moderately damaged
    d. severely damaged
    e. life safety
    f. near collapse

92. "A philosophy quickly developed suggesting that existing buildings be treated differently from new buildings with regard to seismic requirements. First, archaic systems and materials would have to be recognized and incorporated into the expected seismic response, and secondly, due to cost and disruption, seismic design force levels could be smaller. The smaller force levels were rationalized as providing minimum life safety, but not the damage control of new buildings, a technically controversial and unproven concept, but popular. Commonly existing buildings were then designed to _____ of the values of new buildings—a factor that can still be found, either overtly or hidden, in many current codes and standards for existing buildings.".
    a. 60%
    b. 75%
    c. 80%
    d. 85%

93. As the conceptual framework of evaluation and retrofit developed, legal and code requirements were also created. These policies and regulations can be described in three categories: (**Check the three that apply.**)
    a. active
    b. passive
    c. post-earthquake

    d. preventive

    e. retrofit

94. Which of the following statements is true? **Check the two that apply.**

    a. After an earthquake, all buildings should be inspected by a building official.

    b. After an earthquake, all buildings should be inspected and Green-tagged by a building official if they are undamaged.

    c. After an earthquake, damaged buildings should be either Green-tagged or Yellow-tagged.

    d. After an earthquake, a damaged building that creates a public risk and requires immediate mitigation should be Red-tagged.

95. Which of the followings statements is true? **Check the two that apply.**

    a. The principal seismic risk in the United States comes from the existing building stock.

    b. The principal seismic risk in the United States comes from selecting the wrong occupancy category when designing buildings.

    c. The FEMA "yellow book" series are less known to architects.

    d. The FEMA "yellow book" series are less known to engineers.

96. Considering only the benefit-cost ratio of seismic retrofit, which building type is most likely to be retrofitted?

    a. unreinforced masonry (URM) buildings

    b. tilt-up buildings

    c. wood-stud buildings

    d. moment-resisting frame buildings

97. Which of the following is a flexible diaphram? **Check the two that apply.**

    a. metal deck roof over roof joists over wood trusses

    b. plywood over floor joists

    c. cast-in-place concrete floor

    d. pre-cast concrete floor

98. According to FEMA 454, what is the expected lifespan of a building?

    a. 50 years

    b. 75 years

    c. 100 years

    d. 150 years

99. According to FEMA 454, nationally applicable building codes are based on the level of shaking intensity expected at any site once every ___ years (on average).

    a. 50

    b. 100

    c. 300

    d. 500

100. Which of the following statements is not true?
    a. The seismic performance of nonstructural systems is not important since it will not affect life safety.
    b. The seismic performance of nonstructural systems is important since it will affect property loss and/or life safety.
    c. Historically, the seismic performance of nonstructural systems and components has received little attention from designers.
    d. Some investigators have postulated that nonstructural system or component failure may lead to more injury and death in the future than structural failure.

101. Who is responsible for most of the nonstructural seismic design issues for both systems and components?
    a. the architect
    b. the structural engineer
    c. the electrical engineer
    d. the mechanical engineer
    e. the specialty consultant

102. For a single family residence, which of the following should be connected to UFER or other kinds of ground?
    a. electrical lines
    b. cable lines
    c. phone lines
    d. all of the above

103. The permitted wind design procedure includes all of the following EXCEPT
    a. simplified procedure
    b. analytical procedure
    c. wind tunnel procedure
    d. diagram procedure

104. Inland waterways belong to which wind exposure category per ASCE 7?
    a. Exposure A
    b. Exposure B
    c. Exposure C
    d. Exposure D

105. Figure 3.25 shows a reinforced concrete wall. Which of the following statements is correct?
    a. Pier 1 and Pier 2 resist more lateral load than Pier 3 and Pier 4.
    b. Pier 1 and Pier 3 resist more lateral load than Pier 2 and Pier 4.
    c. Pier 1 and Pier 4 resist more lateral load than Pier 2 and Pier 3.
    d. Pier 3 and Pier 4 resist more lateral load than Pier 1 and Pier 2.
    e. Each pier resists the same lateral load.

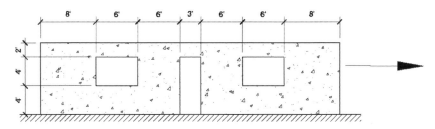

Figure 3.25   Distribution of lateral load for a reinforced concrete wall

106. Figure 3.26 shows a 2-story room addition to a 1-story existing building. Which of the following statements is correct?
    a. Column A will resist more lateral load than the other columns.
    b. Column B will resist more lateral load than the other columns.
    c. Column C will resist more lateral load than the other columns.
    d. Column D will resist more lateral load than the other columns.
    e. Each column resists the same lateral load.

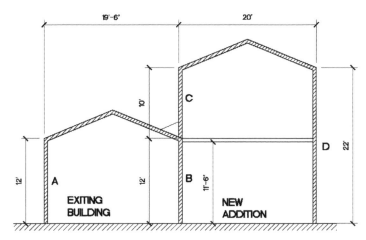

Figure 3.26   Which column will resist more lateral load than the other columns?

107. Which of the following is a minimum slope for a flat roof?
    a. 1/8" per foot (1%)
    b. 1/4" per foot (2%)
    c. 1/2" per foot (4%)
    d. 3/4" per foot (6%)

108. The wind maps referred to by IBC and ASCE for non-hurricane-prone regions are based on wind speed of
    a. a yearly 1% probability of occurrence
    b. a yearly 2% probability of occurrence
    c. a yearly 4% probability of occurrence
    d. a yearly 5% probability of occurrence

109. Diaphragm boundary is a term frequently used in
    a. light-frame construction
    b. heavy timber construction
    c. masonry construction
    d. tilt-up construction

110. Per IBC, "where the live loads for which each floor or portion thereof of a commercial or industrial building is or has been designed to exceed _____, such design live loads shall be conspicuously posted by the owner in that part of each story in which they apply, using durable signs. It shall be unlawful to remove or deface such notices."
    a. 25 psf (1.20 kN/m²)
    b. 50 psf (2.40 kN/m²)
    c. 100 psf (4.80 kN/m²)
    d. 200 psf (9.60 kN/m²)

111. In the United States, which of the following natural disasters cause the most damage to buildings?
    a. wind
    b. earthquake
    c. flood
    d. mountain fire

112. During a field visit, a building inspector asks the contractor to nail roof sheathing to the blocking along the exterior wall. This is to prevent potential damages caused by
    a. wind
    b. earthquake
    c. none of the above
    d. both a and b

113. According to basic wind speed, a 3-second gust is _____ mph for almost all of the continental United States, except the east and west coast.
    a. 60
    b. 80
    c. 90
    d. 120

114. Figure 3.27 shows a 3-story moment-resisting frame with hinged bases resisting lateral loads. Ignoring the dead loads, what force resists the overturning caused by the lateral loads?
    a. shear in the columns
    b. moment at the column bases
    c. tension in one hinge base, and compression in another hinge base
    d. tension in the beams
    e. compression in the beams

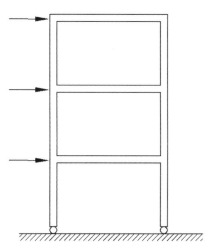

Figure 3.27   Moment-resisting frame with hinged bases

115. Referring to the same 3-story moment-resisting frame as shown in figure 3.27, if we include the dead load, what resists the overturning caused by the lateral loads? **Check the two that apply.**
   a. shear in the columns
   b. weight of the columns
   c. moment at the column bases
   d. tension in one hinge base, and compression in another hinge base
   e. tension in the beams
   f. weight of a column and beams
   g. compression in the beams
   h. tension or compression in one hinge base, and compression in another hinge base

116. Which of the diagrams in figure 3.28 show the deflected shape of a rigid frame with a rigid base for the loading shown?

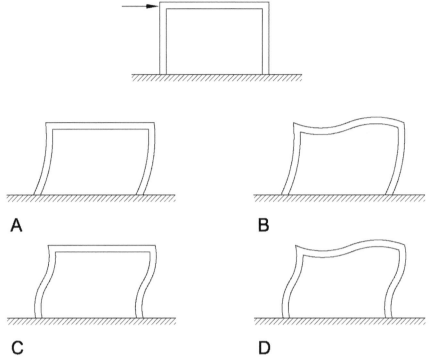

Figure 3.28   Diagram for a rigid frame

117. Which of the diagrams in figure 3.29 shows the deflected shape of a rigid frame with a rigid base for the loading shown?

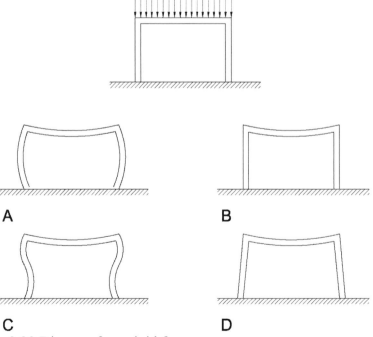

Figure 3.29 Diagram for a rigid frame

118. A horizontal force of 25 kips is applied to a frame with diagonal braces made of cables as shown in figure 3.30. What is the internal tension in each brace?
   a. brace 1 = 0, brace 2 = 35.36 kips
   b. brace 1 = 0, brace 2 = 25 kips
   c. brace 1 = 35.36, brace 2 = 0 kips
   d. brace 1 = 25, brace 2 = 0 kips

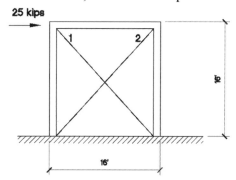

Figure 3.30   Diagram for a frame

119. Figure 3.31 shows a moment-resisting frame. Which of the following statements is correct? **Check the two that apply.**
   a. The bottom of the column is free to rotate.
   b. The bottom of the column is fixed against rotation.
   c. The top of the column is free to rotate.
   d. The top of the column is fixed against rotation.

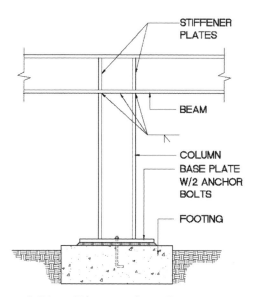

Figure 3.31   Diagram for a frame

120. Which of the diagrams in figure 3.32 show the distribution of lateral forces used in seismic design?

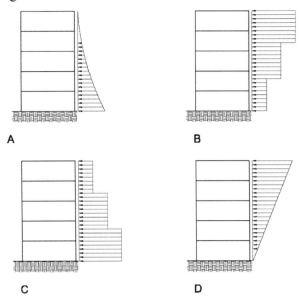

Figure 3.32   Diagram for seismic forces

121. Which of the systems shown in figure 3.33 is stable under lateral forces? **Check the two that apply.**

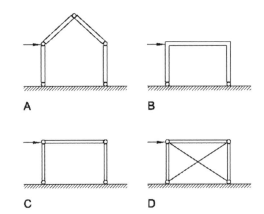

Figure 3.33   Diagram for seismic forces

122. Figure 3.34 shows a building's floor plan. The total wind load in the east-west direction
    a. is greater than in the north-south direction
    b. is the same as in the north-south direction
    c. is smaller than in the north-south direction
    d. can be more or less than in the north-south direction, depending on the gust factor

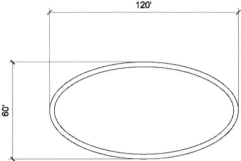

Figure 3.34   A building's floor plan

123. Figure 3.35 shows beam to column connections. Which of the following is part of a moment-resisting frame?
    a. I
    b. II
    c. I and II
    d. Neither I or II

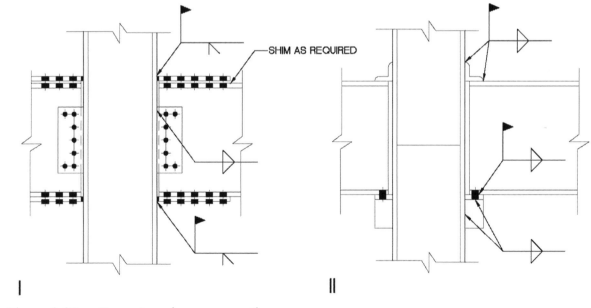

Figure 3.35   Beam to column connections

124.   Figure 3.36 shows a truss. Which of the following shows the reactions at A and B correctly?

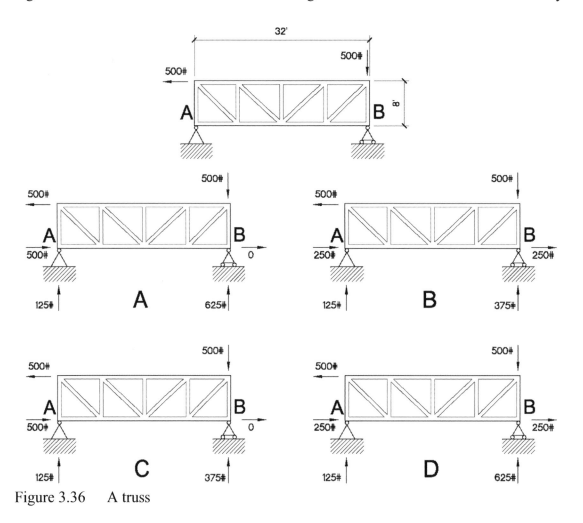

Figure 3.36   A truss

125.   Figure 3.37 shows a shear wall with 500 lb of dead load on each end, and 500 lb of lateral load. What is the net uplift at point 1?
   a. 10,000 lb
   b. 5,000 lb
   c. 2,500 lb
   d. zero

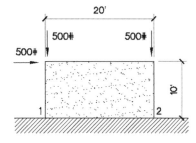

Figure 3.37   A shear wall

B. Mock Exam: SS Graphic Vignette Section

Directions

Use the NCARB software to create a 2-level framing solution over the program areas (figure 3.38).

Program

The same as the NCARB sample vignette.

Warning

You must meet both of the following requirements, OR you will automatically fail:
1. Draw the solution on the right layer.
2. No cantilevers are allowed.

Time

You must complete the vignette within 1 hour

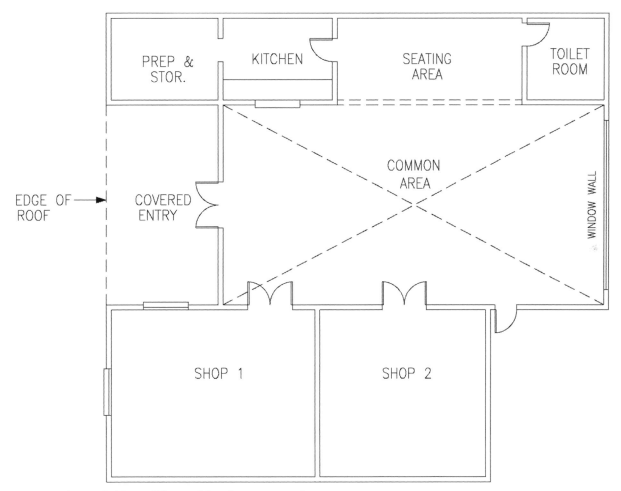

Figure 3.38   SS graphic vignette mock exam

C. Instructions on Installing Alternate DWG Files for Use with NCARB Software

1) **Right click** the **Start** button on the lower left-hand corner of your computer to open your **Windows Explore** (figure 3.39).
2) Go to the folder where you placed the downloaded DWG file. On the top pull-down menu, under **View**, select **Details**. You should see an extension for all the file names, i.e., a dot (.) followed by three letters. The AutoCAD file name for our alternate drawing is **SS Figure 3.38.dwg**; the ".dwg" is the extension (figure 3.40).
3) If you do NOT see an extension for all the file names, see the following instructions. (These directions are for Windows Vista and Window 7, but Windows XP is similar.)
   - **Windows Explore > Organize > Folder and Search Options** (figure 3.41) A menu window will pop up. In that menu, select **View** (figure 3.42).
   - You will see a list with several boxes checked. Scroll down and **uncheck** the box for **Hide extensions for known file types.**
   - Select **Apply to Folders** and a menu window will pop up. Select **Yes** to get out of the **View** menu (figure 3.43).
4) **Windows Explore > Computer > C: Drive > Program Files > NCARB.** Select the folder for the NCARB ARE division that you are working on (figure 3.44). For this instance, choose **C:\Program Files\NCARB\Structural Systems.**
5) Open the folder and you will see files ending in .AUT and .DWG. The .AUT files are the program information and .DWG files are the templates for the practice vignettes. This is important.
6) Make a new folder called **Backup** under the **Structural Systems** folder. Copy all the .AUT files and .DWG files to the **Backup** folder. For SS, we only have one DWG format file: **C2TUT3W1.dwg.**
7) Make a new subfolder called **Alt** in the **Structural Systems** folder. Copy the alternate DWG file(s) (**SS Figure 3.38.dwg**) that you want to use into the Alt folder. Rename the alternate DWG file(s) to match the name of the original NCARB DWG file. For our example, we will rename **SS Figure 3.38.dwg** as **C2TUT3W1.dwg** to match the original NCARB DWG file.
8) Copy the alternate DWG file(s) to the NCARB folder for your vignette (**C:\Program Files\NCARB\Structural Systems**) and overwrite the original NCARB DWG files (figure 3.45).

Note: NCARB practice software ARE 4.0 ONLY works with AutoCAD Release 12 version. If you have a DWG file that is in an AutoCAD Release 13 version or higher, the NCARB practice software ARE 4.0 will NOT work, and you have to save the DWG file as AutoCAD Release 12 version file.

When you save the DWG file AutoCAD Release 12 version file, you may lose some information such as the pen weights, the leader arrow size, etc. However, you can still read the plan and understand the concepts for this exercise.

9) Open the NCARB practice software. You may get an error message that says the DWG has been changed. Just ignore it and click OK.
10) Now you can start to work on your solution using the NCARB software.

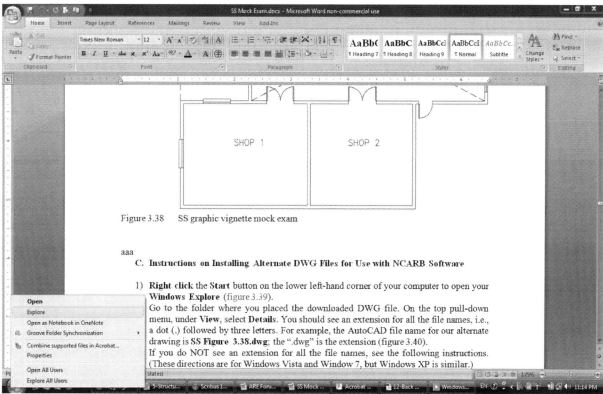

Figure 3.38   SS graphic vignette mock exam

aaa

C. Instructions on Installing Alternate DWG Files for Use with NCARB Software

1) **Right click** the **Start** button on the lower left-hand corner of your computer to open your **Windows Explore** (figure 3.39).
Go to the folder where you placed the downloaded DWG file. On the top pull-down menu, under **View**, select **Details**. You should see an extension for all the file names, i.e., a dot (.) followed by three letters. For example, the AutoCAD file name for our alternate drawing is **SS Figure 3.38.dwg**; the ".dwg" is the extension (figure 3.40).
If you do NOT see an extension for all the file names, see the following instructions. (These directions are for Windows Vista and Window 7, but Windows XP is similar.)

Figure 3.39   **Right click** the **Start** button on the lower left-hand corner of your computer to open **Windows Explore.**

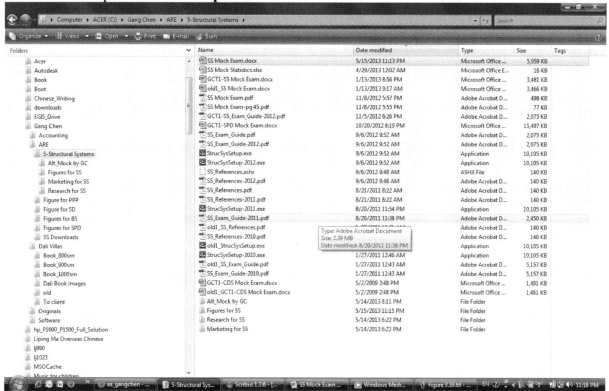

Figure 3.40   You should see an extension for all the file names, i.e., a dot (.) followed by three letters.

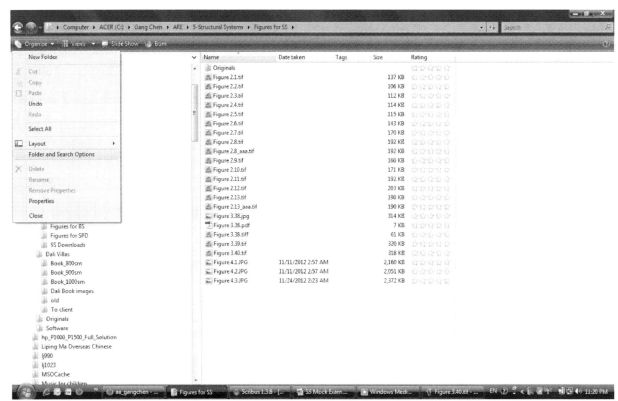

Figure 3.41   **Windows Explore > Organize > Folder and Search Options**

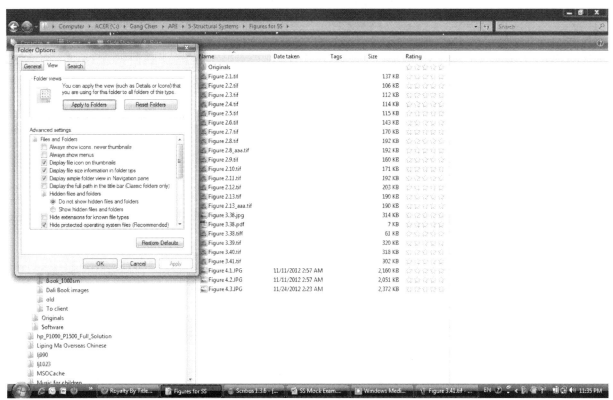

Figure 3.42   A menu window will pop up. In that menu, select **View.**

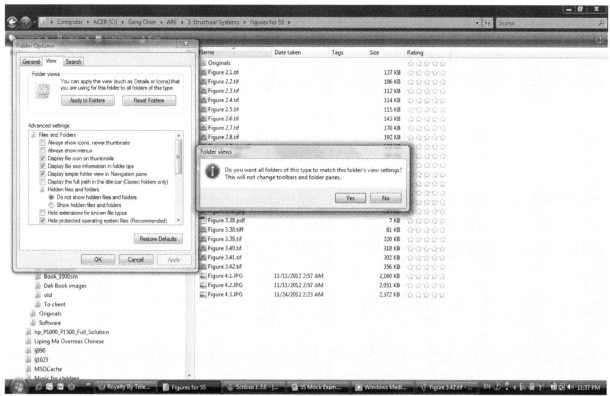

Figure 3.43   Select **Apply to Folders** and a menu window will pop up. Select **Yes** to get out of the **View** menu.

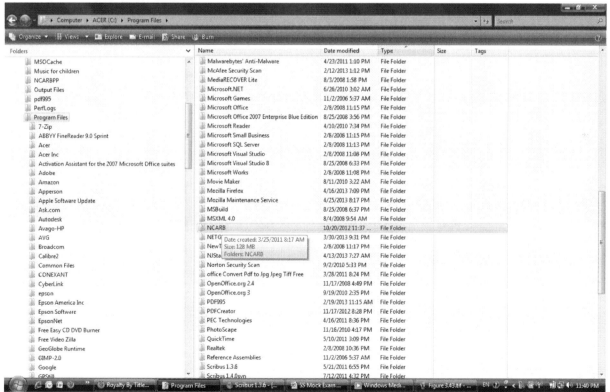

Figure 3.44   **Windows Explore > Computer > C: Drive > Program Files > NCARB.** Select the folder for the NCARB ARE division that you are working on.

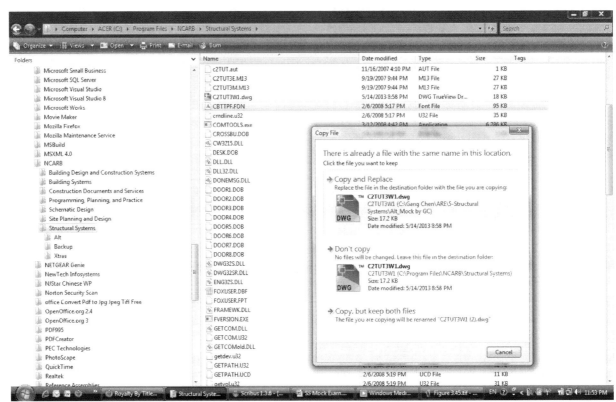

Figure 3.45   Copy the alternate DWG files to the NCARB folder for your vignette and overwrite the original NCARB DWG files.

Chapter Four

ARE Mock Exam Solutions for Structural Systems (SS) Division

A. Mock Exam Answers and Explanations: SS Multiple-Choice (MC) Section

Note: If you answer 60% of the questions correctly, you will pass the MC Section of the exam.

1. Answer: a and c
   Per the ***International Building Code (IBC)***, the following are correct:
   - *IBC* allows live load reduction in most cases.
   - *IBC* does not allow live load reduction for public assembly occupancy with a live load equal to or less than 100 psf.

   The following are incorrect answers:
   - *IBC* allows live load reduction in a few cases.
   - *IBC* does not allow live load reduction for live load equal to or more than 100 psf.

   See the ***International Building Code (IBC)*** at following link:
   http://publicecodes.cyberregs.com/icod/ibc/2006f2/icod_ibc_2006f2_16_par065.htm?bu=IC-P-2006-000001&bu2=IC-P-2006-000019

2. Answer: a
   Dynamic loads are loads that change rapidly.

   Impact loads are suddenly applied loads.

   Dead loads are vertical loads generated by the weight of the building or permanent equipment.

3. Answer: b and d
   Please note we are looking for *incorrect* statements.

   The following are *incorrect* statements, and therefore the *correct* answers:
   - Wind velocity does not affect wind loading.
   - Wind velocity is lower on the 10th floor than on the ground floor.

The following are *correct* statements, and therefore the *incorrect* answers:
- Wind velocity affects wind loading.
- Wind velocity is higher on the 10th floor than on the ground floor

Ground surfaces cause friction and slow the wind, therefore, the wind velocity is lower closer to the ground.

4. Answer: If a load of 10,000 lb is applied to a 10' long, 4x4 Douglas fir no. 2 wood, it will compress <u>0.058 inch or about 1/16</u> inch.

Note:
We intentionally place some questions requiring calculations at the front of the mock exam to train your ability in time management.

In the ARE exams, it may be a good idea to skip any calculation question that requires over 30 seconds of your time; just pick a guess answer, mark it, and come back to calculate it at the end. This way, you have more time to read and answer other easier questions correctly.

Since the modulus of elasticity for Douglas fir no. 2 is 1,700,000 psi, we can calculate the total deformation (change in length) using the following formula:

$$e = \frac{PL}{AE} = \frac{(10,000 \text{ lbf}) (10 \text{ ft}) (12 \text{ in/ft})}{(3.5 \text{ in} \times 3.5 \text{ in}) (1,700,000 \text{ lbf/ in}^2)} = 0.058 \text{ in} = \text{about } 1/16 \text{ in}$$

P = force (lbf) = 10,000 lbf
L = length of column (in) = (10 ft) (12 in/ft) = 120 in
A = cross sectional area (in^2) = (3.5 in x 3.5 in) = 12.25 in^2

Note: The actual dimension of a 4x4 is 3.5" x 3.5", 4x4 is only the nominal dimension.
E = modulus of elasticity (stress/strain = lbf/in^2)

This is an important equation. You need to become very familiar with all the variables, and memorize this equation if possible.

Per the feedback of ARE candidates, there are at **least 20 questions which require calculations** in the real ARE exams. This means you need to be very familiar with some of the most fundamental equations.

The following may help you understand and memorize the equation:

Stress (f) is the force per unit area, and equals to **the total force (P)** divided by the **total area (A)**.
f = P/A

Strain (ε) is the deformation caused by external forces, and it equals to the ratio of the total **length change (e)** to the **original length (L)** of a material.
ε = e/L

The modulus of elasticity (E) is a measure of a structural member's *material* stiffness. It is the ratio of stress to strain.
E = f/ε = (P/A) / (e/L) = PL/Ae

Therefore:
e = PL/AE

5. Answer: Locate the centroid of the hatched area as shown in figure 3.1 and calculate to find that *x* = 9.70".

The centroid is the "center of gravity" of a very thin and flat object or a plane surface. We select X-X axis as the convenient base of the object. The sum of the statical moment of all parts equals the statical moment of the whole object, and we treat the area of the hole in the object as a negative number (figure 4.1):

(6" x 10") (13") + (10" x 4") (5") + (-π x 1²) (13") = x (6" x 10"+ 10" x 4" -π x 1²)

780 + 200 - 40.82 = x (60 + 40 - 3.14)

939.18 = x (96.86)

x = 939.18 / 96.86 = 9.70"

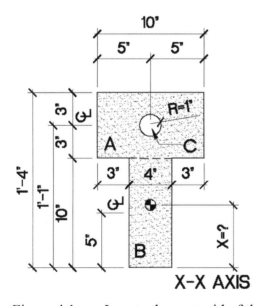

Figure 4.1 Locate the centroid of the hatched area

6. Answer: The moment of inertia for the composite section as shown in figure 3.2 is 2049.33 ft-lb

The moment of inertia (I) is a measure of the bending stiffness of a structural member's cross-section *shape*. It is the ratio of stress to strain.

For a rectangular section, the moment of inertia about a **horizontal** neutral axis is

$$I = \frac{bd^3}{12}$$

b = the width of the rectangular section
d = the height of the rectangular section

The moment of inertia about a **vertical** axis is

$$I = \frac{bd^3}{3}$$

Without special reference, when we use the term, moment of inertia (I), we typically refer to the moment of inertia about a horizontal neutral axis.

We can transfer the moment of inertia of a composite section's parts to a new centroidal axis using the following equation:

$$I_n = I_x + Ad^2$$

The sum of the transferred moment of inertia is the **composite section's overall moment of inertia**.

Now, let us work on the specificities of this problem.
First, find the centroid of the object using a procedure similar to that given for question 5,
(6" x 10") (13") + (10" x 4") (5") = x (6" x 10"+ 10" x 4")

$x = 9.8$"

We can use a table to simplify our calculations.

Area	I_o (in^4)	A (in^2)	d (in)	Ad^2 (in^4)	$I_o + Ad^2$ (in^4)
A	180	60	3.20	614.40	794.40
B	333.33	40	4.80	921.60	1254.93
The moment of inertia for the composite section = sum of ($I_o + Ad^2$) of Areas A and B					2049.33

$$I_o = \frac{bd_o{}^3}{12}$$

d_o = width of the part (A or B)
d = distance from centroidal axis to the axes of each part (Area A or B)

7. Answer: c
    The reactions of the beam shown in figure 3.3 are
    R_1 = 1,300 lbf, R_2 = 4,700 lbf.

    Detailed calculations
    Take the sum of the moment about R_1 to eliminate one of the unknown reactions and simplify the calculation.

    Since the beam is statically determinate, the moment sum of the loads and reactions must equal zero.

    $$(2,000 \text{ lbf}) (4 \text{ ft}) + (3,000 \text{ lbf}) (8 \text{ ft}) + [(500 \text{ plf}) (2 \text{ ft})] (15 \text{ ft}) - (10 \text{ ft}) R_2 = 0$$

    $$R_2 = 4,700 \text{ lbf}$$

    The sum of the loads must equal the reactions.
    $$2,000 \text{ lbf} + 3,000 \text{ lbf} + (500 \text{ plf}) (2 \text{ ft}) = R_1 + R_2$$

    $$6,000 \text{ lbf} = R_1 + 4,700 \text{ lbf}$$

    $$R_1 = 6,000 \text{ lbf} - 4,700 \text{ lbf} = 1,300 \text{ lbf}$$

8. Answer: b
    Figure 3.4 shows a shear diagram of concentrated loads.

    Upward forces are positive, and downward forces are negative. When there are no intervening loads between two concentrated loads, the portion of the shear diagram is a horizontal line. An intervening load causes a shear diagram to change abruptly vertically.

    A shear diagram of uniformly distributed loads is a sloped line.

9. Answer: d
    The moment diagram of uniformly distributed loads is a parabolic shape.

    The moment diagram of concentrated loads is composed of triangles.

10. Answer: a
    The location where the line of moment diagram crosses zero is the **inflection point**. This is where the rebar in concrete beams are bent, and switch location from the *bottom* of the beam to the *top* of the beam. This accounts for the change from *positive* moment to *negative* moment.

    Deflection is the change in vertical position of a beam due to load.

    Zero point and balance point are **distracters**.

11. Answer: b, c, and e
The following are methods to determine the forces in truss members:
- method of joints
- method of sections
- graphic method

The following are **distracters**:
- method of moment
- method of analysis

12. Answer: c
A **standard penetration test (SPT)** tests the density and consistency of soil. A 140-lbf hammer falls 30 inches onto a 2-inch diameter sample which drives it into the bottom of a borehole. The number of blows needed to drive the cylinder 12 inches is recorded.

The following are **distracters**:
- ability of water to penetrate soil
- ability of water to penetrate retaining walls
- density and consistency of slab

13. Answer: b
USCS stands for **Unified Soils Classification System**. It divides soil into major divisions and subdivisions per grain size and physical characteristics.

The following are **distracters**:
- Unified Standard Classification System
- Unified States Concrete Society
- none of the above

14. Answer: a, c, and d.
There are three types of retaining walls.
- **Gravity walls** resist forces by their own weight (gravity) and soil friction.
- **Cantilever walls** are reinforced concrete walls, which resist forces by their own weight and by the weight of the soil on the heel of the base slab.
- **Counterfort walls** are similar to cantilever walls, but with counterfort reinforcement added at spacing roughly equal to half the height of the wall.

The following are incorrect answers:
- brace wall (It can be a retaining wall, a parapet wall, or another wall.)
- CMU wall (It can be a retaining wall, a parapet wall, or another wall.)

15. Answer: b

Figure 3.5 shows a **shear plate connector**. It has a flat plate with a flange extending from the face of the plate.

The following are incorrect answers:

- **split ring connector** (It has a cut-through ring to form a tongue and slot, but no flat plate.)
- double shear connector
- multiple shear connector

Figure 3.5 shows a single shear connector, not a double shear connector or a multiple shear connector.

16. Answer: a

Figure 3.6 shows a field weld; it is also welded all around.

The following are incorrect answers:

- flare weld
- bevel weld
- square weld

As an architect, you need to be able to read and understand the weld symbols. See the following links for more information:

http://en.wikipedia.org/wiki/File:Elements_of_a_welding_symbol.PNG
http://www.typesofwelding.net/weld_symbol.html
http://www.typesofwelding.net/weld_joints_symbol.html

You can also find information on weld joints and weld symbols in the *AISC Manual of Steel Construction*.

17. Answer: a and b

The concerns for a roof without sufficient slope include the following:

- **Ponding** is caused by inadequate drainage.
- **Deflection is** caused by accumulation of water on a roof.

The following are incorrect answers:

- **Inflection** is a concept related to moment diagrams. The location where the line of the moment diagram crosses zero is the **inflection point**.
- **Failure in an earthquake** has no direct relationship with a sufficient roof slope.

18. Answer: c

Aluminum conduit cannot be directly embedded in a concrete foundation or slab. Aluminum can have reactions with concrete or electrolytic action with the steel reinforcement of concrete.

The following are incorrect answers:

- ABS pipe (They can be embedded in regular concrete without any problem.)
- PVC pipe (They can be embedded in regular concrete without any problem.)
- copper conduit (They can be embedded in regular concrete without any problem. However, copper should be protected when it comes into contact with concrete mixtures containing components high in sulfur like fly-ash and cinders, which can create an acid that is highly corrosive to most metals including copper. Copper should not be in direct contact with the steel reinforcement of concrete in order to avoid electrolytic action.

19. Answer: c

The structural steel shown in figure 3.7 is a "S" shape.

Look through the *AISC Manual of Steel Construction* and become familiar with the common steel shapes.

20. Answer: a

A steel column with a slenderness ratio of 32 will fail first by buckling under loads.

The greater the slenderness ratio, the greater the tendency for a column to fail by buckling first.

21. Answer: d

If end conditions are the only differences, answer "d" of the column list below has the highest slenderness ratio.

$$\text{slenderness ratio} = \frac{Kl}{R}$$

K - a value to modify the unbraced length of a column
l – the length of a column
r - **radius of gyration**

If we arrange the slenderness ratio of the columns from highest to lowest, the order of the columns is d > e > c > b > a. See the table "K-Values for Various End Condition" in the *AISC Manual of Steel Construction.*

Here is the list of the columns:
a. steel column fixed in both rotation and translation at both ends
b. steel column fixed in both rotation and translation at one end, but free to rotate and fixed in translation at the other
c. steel column fixed in both rotation and translation at one end, but free to translate and fixed in rotation at the other
d. steel column fixed in both rotation and translation at one end, but free to rotate and translate at the other
e. steel column free to rotate but fixed in translation at both ends

22. Answer: b

$$\textbf{slenderness ratio} = \frac{\textbf{K}\textbf{\textit{l}}}{\textbf{\textit{r}}} = \frac{(0.65)\,(16\text{ ft})\,(12\text{ in/ft})}{3.13\text{ in}} = 39.87$$

Note:
Make sure you convert the column length to inches and use the radius of gyration with the least value.

23. Answer: c

UFER is an item for electrical work. It is an electrical earth grounding method, and uses a concrete-encased electrode to improve grounding in dry areas. It is used in construction of concrete foundations.

The NEC refers to this type of ground as a "Concrete Encased Electrode" (CEE) instead of a UFER ground.

24. Answer: a, b, and d

A contractor is working on a single-family home project, the following subcontractors must complete the underground work before s/he can pour the concrete slab and foundation:
- framing subcontractor (placing anchor bolts and hold-downs before the concrete is poured)
- electrical subcontractor
- plumbing subcontractor

The plaster subcontractor and mechanical subcontractor typically start their portion of the work AFTER the slab and foundation is completed. Therefore these answers are incorrect

25. Answer: d

First, convert the loads to load per linear foot of joist:
Since joists are spaced 4 ft on center, dead load = (30 psf) (4 ft) = 120 plf, live load = (50 psf) (4 ft) = 200 plf, and the total load = 120 plf + 200 plf = 320 plf.

Second, look at table 3.1 across the 36 ft span row:
20K10 can support 369 plf of total load, and 193 plf of live load. (Live load is typically shown right below the total).
22K9 can support 344 plf of total load, and 201 plf of live load.

Both 20K10 and 22K9 can support the required loads, but 20K10 weighs 12.2 plf, while 22K9 weighs 11.3 plf. So, 22K9 is the most economical section to use.

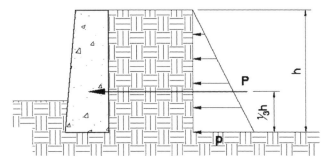

Figure 4.2   The pressure at the bottom of a retaining wall

26. Answer: b. 360 psf
    See figure 4.2. The pressure at the bottom of the wall is P = (4 ft) (45 psf/ft) = 180 psf.

The total horizontal load P = $\dfrac{\mathbf{ph}}{\mathbf{2}}$ = $\dfrac{(180 \text{ psf}) (4 \text{ ft})}{2}$ = 360 psf

The total load can be shown as acting on the centroid of a triangle, 1/3 the distance from the base.

27. Answer: b and c
    Because this is a simply supported beam, the section A-A shown in figure 3.8 has
    • compression stresses in the top half of the beam, and
    • tension stresses in the bottom half of the beam.

28. Answer: a and d
    Because this is a cantilever beam, section A-A shown in figure 3.9 has
    • tension stresses in the top half of the beam, and
    • compression stresses in the bottom half of the beam.

    Figure 3.9 is a very basic but important diagram, you really need to know it like the back of your hand.

29. Answer: a
    Figure 3.10 shows a simple truss. Using the method of joints, the force in member AB (the member between joints A and B) is 10,009 lbf.

    Our step-by-step solution is as follows.
    1) Find the **reactions.**
       Since all forces are symmetric, R_A = R_F = ½ (3,000 lbf + 6,000 lbf + 3,000 lbf) = 6,000 lbf.

    2) We presume F_{AB} is a compression force (pointing toward the joint), and F_{AD} is a tension force (pointing away from the joint), and draw joint A as a **free body diagram**.

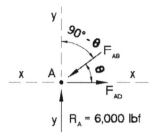

Figure 4.3   A free body diagram for Joint A

3) Find the angle between member AB and AD.

$$\tan \theta = \frac{12 \text{ ft}}{16 \text{ft}}$$

$$\theta = 36.87°$$

The complementary angle of $\theta = 90° - 36.87° = 53.13°$

4) The sum of all vertical forces equal zero. Since F_{AB} is acting downward, it is negative:

$$R_A - F_{AB} (\cos 53.13°) = 6{,}000 \text{ lbf} - F_{AB} = 0$$

$$F_{AB} = \frac{6{,}000 \text{ lbf}}{\cos 53.13°} = 10{,}000 \text{ lbf}$$

Note:
Sometimes, you can tilt the x- and y- axes to simplify the calculations. This is an important technique for the method of joints (figure 4.4).

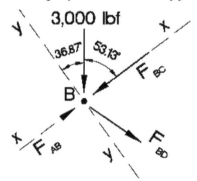

Figure 4.4   A free body diagram for Joint B

30. Answer: b
    For the same simple truss in figure 3.10, using the method of sections, the force in member BC is 7,500 lbf.

    Our step-by-step solution is as follows. (figure 4.5)
1) Use the same procedure in question 29 to find the reactions.
    Since all forces are symmetric, $R_A = R_F = \frac{1}{2}$ (3,000 lbf + 6,000 lbf + 3,000 lbf) = 6,000 lbf

2) Find the moment arm of BC.
    Per calculations in question 29, $\theta = 36.87°$ and the complementary angle is $53.13°$. The difference of the angles = $53.13° - 36.87° = 16.26°$.

$$\cos 16.26° = \frac{H}{10\ ft}$$

    h = (10 ft) (cos 16.26°) = 9.60 ft

3) Take moments about point D to eliminate the two unknowns (F_{BD} and F_{AD}).
    Note:
    Because F_{BD} and F_{AD} pass through joint D, their moment arms equal zero, and therefore their moments equal zero as well. This is an important technique when using the method of sections.

    By convention, moments acting clockwise are positive, and moments acting counterclockwise are negative. The sum of the moments about point D equal zero:

    (6,000 lbf) (16 ft) - (3,000 lbf) (8 ft) – F_{Bc} (9.60 ft) = 0

    $F_{BC} = 7,500$ lbf

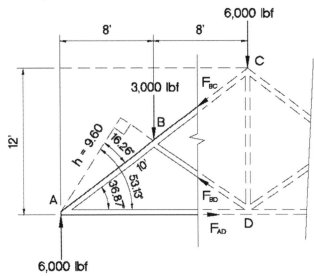

Figure 4.5   Method of sections

31. Answer: a and c

A contractor is working on a wood frame residential project with a slab on grade and continuous footing. The following statements are correct:

- The hold-downs bolts should be placed before pouring the concrete footing and slab.
- The anchor bolts should be placed before pouring the concrete footing and slab.

If the contractor places anchor bolts and hold-downs bolts after pouring the concrete footing and slab, s/he has to spend a lot of time and effort to drill the concrete, and the bonding between the hold-own and concrete may not be strong enough.

Ideally, the anchor bolts and hold-downs bolts should be placed by the framing contractor to avoid problems during framing later on.

32. Answer: c

The architect should ask the structural engineer to change the plans to use 4,500 psi concrete, but note on the structural plans: "4,500 psi concrete is not required by structural design, special inspection is not required."

The following are incorrect answers:

- Ask the structural engineer to change the plans to use 4,500 psi concrete. (This is not 100% correct because concrete with a compressive strength of 3,000 psi or higher typically requires inspection except in this situation. The structural engineer should note on the structural plans: "4,500 psi concrete is not required by structural design, special inspection is not required." This will prevent the unnecessary extra time and cost of special inspection.)
- Immediately respond to the plan check correction and refuse to use the 4,500 psi concrete because it is not required by the structural design.
- Ask the structural engineer to change the plans to use 4,500 psi concrete, and note the structural plans "special inspection is required."

Another option not listed in the original choices is to ask the owner to do a soils report to investigate the existence of sulfate at the project site. If no sulfate is present at the site, then the architect can respond to the plan check correction and refuse to use the 4,500 psi concrete.

33. Answer: b and d

The following structural members are typically Douglas fir #1 instead of #2:

- 4x6 columns
- 4x8 beams

In fact, all columns and beams are typically Douglas fir #1 instead of #2.

The following structural members are typically Douglas fir #2 instead of #1:

- 2x10 floor joists
- 4x6 window and door headers

34. Answer: b

Figure 3.11 shows an anchor bolt. An anchor bolt has a "J" shape. As an architect, you need to be able to identify some very basic structural items.

The following are incorrect answers:
- lag screw
- hold-down bolt
- base plate screw

35. Answer: c

Figure 3.12 shows a hold-down bolt. A hold-down bolt has a "S" shape at the bottom.

The following are incorrect answers:
- lag screw
- anchor bolt
- base plate screw

36. Answer: Number 6 rebar is $6 \times (1/8") = 6/8 = 3/4$ inch in diameter. Similarly, a number 5 rebar is equal to $5 \times (1/8")$ or 5/8 inch in diameter. A rebar's diameter equals its number times 1/8 in.

37. Answer: c

Plasticizers are used to reduce the amount of water needed for mixing concrete.

The following are incorrect answers:
- **Air-entraining agents** are used to form small dispersed bubbles in the concrete.
- **Accelerators** are used to speed up hydration of concrete to achieve strength faster.
- **Fine aggregates** are sand and small crushed rocks. They are part of the normal components of any concrete.

38. Answer: c

Figure 3.13 shows a hold-down, a very common structural component. An architect should be able to identify it.
The following are incorrect answers:
- column base (typically a thin metal plate bent in a U-shape)
- strap (typically a long and thin metal plate with holes for nailing)
- structural plate

39. Answer: c

The diameter of #3 rebar is 3/8". The diameter of a rebar = its number x 1/8"
Similarly, the diameter of #5 rebar is 5/8".

40. Answer: a and d

6x6-10/10 and 6x6-W1.4x1.4 are two different designations for the same welded wire fabric (**WWF**) in English System. Welded wire fabric (**WWF**) is also called welded wire mesh (**WWM**).

Welded wire fabric is a common reinforcement for concrete sidewalks. Becoming familiar with this material will help you in real architectural practice also.

6x6-10/10 is the designation from an old system. 6x6 means the grid of the welded wire fabric is 6 inches x 6 inches, and 10/10 means the wire sizes are both 10 gauges. This old system is still in use.

6x6-W1.4x1.4 is the designation from a new system. 6x6 means the grid of the welded wire fabric is 6 inches x 6 inches, W means it's smooth, and 1.4 means the cross-sectional wire area is 0.014 square inches. If the letter is D instead of W, it means the welded wire fabric is deformed.

6x6-10/10 or 6x6-W1.4x1.4 is designated as 152x152-MW9.1/MW9.1 in the Metric System. The number 152 means 152 mm, and the number 9.1 means 9.1 mm^2.

Again, 6x6-10/10 in the old system is equivalent to 6x6-W1.4x1.4 in the new English System. Many people want to know the detailed explanation. For those who are interested in the detailed information, here it is. You just need to look through this, and do not need to memorize.

The wire sizes are the same in both systems but the old system gives the diameter of the wires which is 10 gauges (or 0.1350") and the new system gives the cross sectional area of the wire, 0.014 square inches. 10-gauge wire is 0.1350" in diameter and has a cross sectional area of 0.0143 square inches, and we can round it to 0.014 square inches.

Note:

The radius of 10-gauge wire =diameter/2 = 0.1350"/2 = 0.0675"

The cross sectional area of 10-gauge wire is:
A = π r^2 = 3.14 x (0.0675)^2 = 0.0143 square inches, and we can round it to 0.014 square inches

The grid size is:
6 inches = 6 x 25.4 mm = 152.40 mm, and we can round it to 152 mm

So, the 6x6 grid size in the new English System is equivalent to the 152x152 in the old system.

41. Answer: a
    Diagram A shows the correct placement of rebar for the concrete beams of a gas station canopy supported by concrete columns.

    Betweens the columns, there is positive moment in the beam and the rebar is placed close to the bottom.

    At the columns and at the cantilever portions of the beam, there is negative moment in the beam and the rebar is bent and placed close to the top.

42. Answer: c
    The gradient height for *metropolitan areas* is approximately 1,500 ft.

    Like water or any other fluid, wind speed is reduced when it is near or in contact with other surfaces. Therefore, the closer to ground, the more wind speed reduces. **Gradient height** is the height above which the friction of ground surfaces and obstacles on the ground do not cause wind speed reduction.

    The gradient height for *open country areas* is approximately 900 ft. The gradient height for *suburban areas* is approximately 1,200 ft.

43. Answer: b
    Using Method II, diagram B shows the correct direction of wind pressure.

    Method II is the normal force method. It assumes that wind pressure acts simultaneously on all exterior surfaces. The windward walls have positive pressures acting toward the surfaces. The roof and leeward walls have negative pressures acting away from the surfaces.

    Most of the questions of the ARE SS division test your understanding of basic concepts. This question is a good example to demonstrate this fact.

    In real practice, you also need to know basic structural concepts like the direction of wind pressures, but you will NOT need to do detailed calculations. The detailed calculations are part of your structural engineers' job.

    NCARB expects the same knowledge since you need to be able to function as an average architect after you pass the ARE exams.

44. Answer: a
    Diagram A is a correct depiction of wind loads on a diaphragm.

    The windward edge of the floor/roof has a compression chord force, and the leeward edge of the floor/roof has a tension chord force.

45. Answer: a

In reference to figure 3.17, the brace at location A is in tension, and the brace at location B is not stressed.

If the wind direction reverses, the brace at location B is in tension, and the brace at location A is not stressed.

K-braces are often placed at the center bay and act as vertical trusses cantilevered out of the ground.

46. Answer: c

Site Class A, B, C, D, E, or F in Chapter 16 of IBC is based on soil properties.

See related section of IBC Chapter 16 at the following link:
http://publicecodes.cyberregs.com/icod/ibc/2006f2/icod_ibc_2006f2_16_par120.htm

47. Answer: b

For a hospital, the Important Factor (I Factor) for **snow** is **1.2**, For earthquakes it is 1.5, and for wind it is 1.15.
You do not have to remember all of these numbers for every building type, but you need to be familiar with them. Since hospitals belong to one of the most important occupancy groups, you need to know them well.

48. Answer: a

Camber is the rise in the middle of a beam. The purpose of camber in a beam is to compensate for deflection.

The following are incorrect answers:
- to compensate for torsion
- to compensate for tension (Deflection is caused by both tension at the bottom of a beam and compression at the top, so tension is an incomplete answer.)
- to compensate for compression (Deflection is caused by both tension at the bottom of a beam and compression at the top, so compression is an incomplete answer.)

49. Answer: a

A soft story is most likely to occur at the first floor of a building.

Most users like to have an open first floor with fewer columns or shear walls, or a first floor with greater height. This makes the first floor weaker than other floors, and is termed a **soft story**. The first floor (or grade floor) also has the largest lateral loads. These combined factors make the first floor more vulnerable to lateral forces.

50. Answer: The section modulus for the geometric section shown in figure 3.18 is <u>94.78</u> in³.

If you really understand the formulas given in the question, it is very easy to come up with the solution. Here is our step-by-step solution.

Note: *The formula I = bd³/12 is for **rectangular** sections ONLY, and is for the moment of inertia about the centroidal axis **parallel to the base**.*

1) We can use the formula $I = bd^3/12$ to calculate the moment of inertia for the big 8"x12" rectangle first:
   $I_b = bd^3/12 = 8 \times 12^3/12 = 1152$ in⁴

2) We can use the formula $I = bd^3/12$ to calculate the moment of inertia for the two small 3.5"x10" rectangles as shown with a hatched pattern in figure 4.6.
   $I_s = bd^3/12 = 3.5 \times 10^3/12 = 291.67$ in⁴

3) To get the moment of inertia for the geometric section shown in figure 3.18, we can subtract the moment of inertia for the two small rectangles from the big rectangle.
   $I = I_b - 2I_s = 1152$ in⁴ $- (2 \times 291.67$ in⁴$) = 568.67$ in⁴

4) To get the section modulus (S) for the geometric section shown in figure 3.18
   $S = I/C = 568.67$ in⁴$/6$ in $= 94.78$ in³

   Note: *C is the distance from the neutral axis (centroidal axis) to the outmost part of the section (extreme fiber). It equals 6" as shown in figure 4.6.*

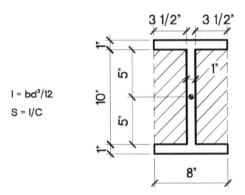

Figure 4.6   Section modulus

51. Answer: The live load for the floor supported by the column as shown in figure 3.19 is <u>75</u> kips (kg).

Each side of the column supports half a bay of the floor (figure 4.7). Therefore, the area supported by the column is: 30' x 50' = 1,500 sf.

The live load for the floor supported by the column as shown in figure 3.19 is:
1,500 sf x 50 psf = 75,000p = 75 kips

All the other formulas and data shown in figure 3.19 are simply **distracters** used to confuse. They are NOT needed for the solution. This is a technique actually used by NCARB for the real ARE exams.

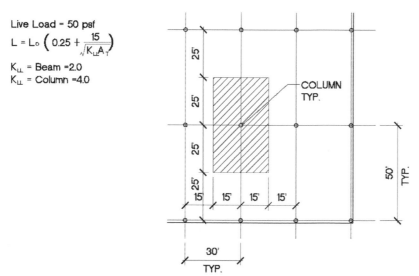

Figure 4.7   The live load for the floor supported by the column

52. Answer: c

The tendency of a solid material to move slowly or deform permanently under the influence of sustained stress is called **creep** or **deformation**.

The following are incorrect answers and used as **distracters**:
- shrinkage
- temperature expansion
- contraction

53. Answer: b

When the ratio of a slab's length (long direction to short direction) is greater than 2:1 it can be considered a one-way slab.

A one-way slab typically only has moment-resisting reinforcement along its short-direction. The moment along its long-direction is very small and can be omitted.

54. Answer: d

The minimum increase in live load due to elevator impact is 100 percent.

According to IBC, Section 1607.8.2, Machinery:

"For the purpose of design, the weight of machinery and moving loads shall be increased as follows to allow for impact: (1) elevator machinery, 100 percent; (2) light machinery, shaft- or motor-driven, 20 percent; (3) reciprocating machinery or power-driven units, 50 percent; (4) hangers for floors or balconies, 33 percent. Percentages shall be increased where specified by the manufacturer."

By the way, item (4) above is also the answer for NCARB sample question #16.

See this related section of IBC for FREE at the following link:
http://publicecodes.cyberregs.com/icod/ibc/2006f2/icod_ibc_2006f2_16_par064.htm?bu2 =undefined

55. Answer: b
B is an eccentrically braced frame (EBF).

Braced frames are designed to have its members work in tension and compression, similar to a truss. Braced frames often have steel members. Most braced frames are **concentric braced frames**, which means that where their members intersect at a node, the centroid of each member passes through the same point.

Concentric braced frames are either **ordinary concentric braced frames (OCBF)** or **special concentric braced frames (SCBF)**. OCBF are often used in areas of low seismic risk.

Special concentric braced frames (SCBF) are also called **eccentrically braced frames (EBF)**. The work points of the diagonal members of an EBF are moved so that the diagonal members are connected to a beam a short distance from the "node" where the beam and column intersect (figure 3.20, Diagram B).

For EBF, there is moment in a beam due to lateral forces. EBF are used in areas of high seismic risk.

In figure 3.20, Diagram A shows an ordinary concentric braced frame (OCBF). Diagram C and D show regular **moment-resistant framing (MRF).**

56. Answer: d
For the flexible diaphragm shown in figure 3.21, the shear at $F_2 = 30$ kips.

A **flexible diaphragm** is one with a maximum lateral deformation more than two times the average story drift of the same story. For a flexible diaphragm, the lateral load and the related shear are distributed per the tributary area. See figure 4.8.

If we assume the length of the building is "L," then the tributary area for F_2 is (shown with hatched pattern in figure 4.8)
$L(15' + 15') = 30L$

The overall area of the building is
$L(30' + 30') = 60L$

The shear at $F_2 = 60$ kips $(30L/60L) = 30$ kips.

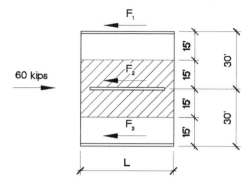

Figure 4.8   A flexible diaphragm

57. Answer: b
If all floors have the same area and similar layout, base isolation is the most effective for a six-story building.

Base isolation is most effective for low-rise buildings. A one-story building would not develop enough base shear to make isolators necessary. The cost of using base isolation for high-rise buildings is prohibitive. High-rise buildings are also more subject to overturning forces, which the base isolators are not effective in dealing with.

58. Answer: a, b, and d

$$\text{slenderness ration} = \frac{Kl}{R}$$

K - a value to modified the unbraced length of a column
l – the length of a column
r - **radius of gyration**

The following affect the slenderness ratio of a column.
• End-constraints affect K and also the slenderness ratio.
• Size of the member affects r and also the slenderness ratio.
• The bigger the radius of gyration, the smaller the slenderness ratio.

The following do NOT affect the slenderness ratio of a column.
• Bracing helps to prevent buckling of a column, but does not affect the slenderness ratio.
• Rotating the column has no effect.

59. Answer: a, d, and e

If an apartment building fails in an earthquake, the following people are likely to have **primary** legal responsibilities:

- the owner
- the contractor (The contractor may have done something wrong in the field which caused the building to fail.)
- the architect (The architect is the "general contractor" of all design professionals; the structural engineer is typically under the architect's contract.)

The following people may have some legal responsibilities, but not primary legal responsibilities:

- the apartment manager (They are typically hired by the owner, and work for the owner. His/her responsibilities are not as primary when compared to those of the owner.)
- the building official (Building officials check the plans, but legal responsibilities remain with the design professionals.)
- the structural engineer (The structural engineer is typically under the architect's contract, but the architect can sue the structural engineer to recover damages in the event of a lawsuit ending in the architect's loss due to negligence by the engineer.)
- Soils engineer (The soils engineer is typically under the owner's contract.)

60. Answer: b

If resisted only by gravity forces, the factor of safety against overturning for a concrete shear wall as shown in figure 3.22 is 6.91.

1) **Calculate the overturning moment (OM)**
   You must use the bottom of the footing to calculate and check the factor of safety (figure 4.9).
   OM = Force x Distance = 30 kips x (30 ft + 3 ft) = 990 kip-ft

2) **Calculate the stabilizing moment (SM)**
   The question does not mention if the 300k of dead load includes the weight of the concrete wall and footing.

   We can try to calculate it both ways.
   - **Assume the 300 kips of dead load (DL) includes the weight of the concrete wall and footing.**
     The distance to the pivot point A is 12.5'.
     SM = DL x Distance = 300 kips x 12.5' = 3,750 kip-ft
     Factor of safety = SM/OM = (3,750 kip-ft)/(990 kip-ft) = 3.79

   - **Assume the 300 kips of dead load (DL) does NOT include the weight of the concrete wall and footing.**
     The weight of the concrete wall = (20' x 30' x 2') x 0.15 kips/ft^3 = 180 kips
     The weight of the concrete footing = (25' x 3' x 6') x 0.15 kips/ft^3 = 67.5 kips

     The distance to the pivot point A is 12.5'.

SM = (DL + The weight of the concrete wall + The weight of the concrete footing) x Distance = (300 kips +180 kips + 67.5 kips) x 12.5' = 6,843.75 kip-ft
Factor of safety = SM/OM = (6,843.75 kip-ft)/(990 kip-ft) = 6.91

3) 6.91 is the correct answer since it matches choice "b" and 3.79 is not an option.

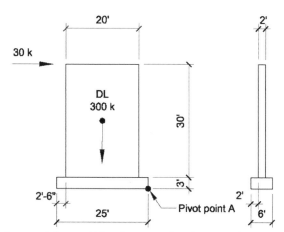

Figure 4.9   The factor of safety against overturning for a concrete shear wall

61. Answer: b and e
    Amongst the choices, the following connections have the least structural value.
    - **High-strength mortar** does not transfer force.
    - **Stacked bond** brick is laid with all vertical joints continuously aligned, the bond between the bricks is very weak.

    The following connections have better structural value:
    - **Bond beams** have steel reinforcement and are very strong.
    - **Metal plate strap anchors** are strong metal parts.
    - **Running bond** brick is laid with joints that fall in the middle of the brick below. It is stronger than stacked bond.
    - **Steel dowels** are very strong metal reinforcement.

    As an architect, you need to be familiar with brick patterns and their features.

62. Answer: d
    The abrupt changes in ground shear distribution at the base of a sufficiently tall structure when subject to critical lateral displacement is called **P-Delta effect**.

    P-Delta effect also includes abrupt changes in the axial force distribution, and/or overturning moment at the base of a sufficiently tall structural component or structure when it is subject to critical lateral displacement.

The following are **distracters**:
- moment shift
- shear interruption
- shear distribution
- overturning moment

63. Answer: d
Per a soils report, the soil bearing capacity is 3,000 psf [143 500 N/m2]. If the applied load is 60,000 lbs [265 kN] for a column, the area of the footing is:

60,000/3,000 = 20 sf

Since the pad footing is square, the width of the pad footing is the square root of 20 sf, or about 4'-6".

64. Answer: d
The minimum concrete coverage for rebar in a **spread footing** is 3" because the footing is in contact with ground. The minimum concrete coverage for #11 or smaller rebar in a **column** or **beam** not in contact with the ground or exposed to weather is 1 ½".

As an architect, you need to know these basic dimensions in order to coordinate with your consultants' work.

65. Answer: c
Pay attention to the word "NOT."
A **hinged frame** is NOT an effective structural system to resist lateral loads because it has flexible joints.

Shear walls and **frames with rigid joints** are effective to resist lateral loads. Each of the following frames/systems have rigid joints:
- frame tube
- portal frame
- moment-resisting frame

66. Answer: a and c
The following are methods used to determine wind loads on a structure (figure 4.10):
- **Normal force method (also called ASCE 7 Method II, the Analytical Procedure)** assumes wind pressures act simultaneously normal (perpendicular) to all exterior surfaces. For the leeward side, the height is taken at the mean roof height and the pressure is considered constant for the full height of the building. The normal force method can be used for any structure, and is the ONLY method for gabled frames with rigid joints.

Note: ASCE means American Society of Civil Engineers

- **Projected area method (also called ASCE 7 Method I, Simplified Method)** assumes that horizontal pressures act on the complete vertical projected area of the building and that vertical pressure acts simultaneously on the complete horizontal projected area. The projected area method can be used for any structure less than 200 feet high EXCEPT for gabled frames with rigid joints.

The following are invented terms used as distracters:
- lateral force method
- diagram method

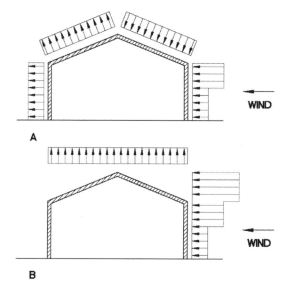

Figure 4.10   Methods to determine wind loads

67. Answer: a

In regard to wind load design, the categories B, C, and D are based on **surface roughness**.

Per IBC, Section 1609.4.2 the surface roughness categories are as follows:
"Surface Roughness B. Urban and suburban areas, wooded areas or other terrain with numerous closely spaced obstructions having the size of single-family dwellings or larger.

Surface Roughness C. Open terrain with scattered obstructions having heights generally less than 30 feet (9144 mm). This category includes flat open country, grasslands, and all water surfaces in hurricane-prone regions.

Surface Roughness D. Flat, unobstructed areas and water surfaces outside hurricane-prone regions. This category includes smooth mud flats, salt flats and unbroken ice."

See IBC Section 1609.4.2 Surface Roughness Categories, for FREE at following link:
http://publicecodes.cyberregs.com/icod/ibc/2006f2/icod_ibc_2006f2_16_par099.htm

68. Answer: a

Diagram A is the correct diagram of wind pressure for a 12-foot high building with a 30-foot square plan.

There is a formula and several tables you can use to calculate the exact wind loads.

You do not need to memorize the complicated process of calculating wind loads, but you do need to have a general idea of the directions and distribution of the wind pressures as shown in the diagram and know the basic concepts very well.

For those who are interested in the detailed information, here it is. You just need to look through this, and do not need to memorize.

Per **American Society of Civil Engineers (ASCE)** 7, the following formula can determine wind pressure (P)

$$P = qGC_p - q_i(GC_{pi})$$

If a building is symmetric, the previous formula can be simplified as:

$$P = qGC_p$$

G: This factor takes into account both aerodynamic and atmosphere effect.
G = 0.85 for rigid structures.

C$_p$: This factor takes into account the different effects of the wind on different parts of the building. It can be *negative* when there is negative pressure on the leeward side or roof. You can look it up in *Minimum Design Loads for Buildings* published by ASCE.

Suppose the horizontal distance from the windward edge is *d*, and building height is *h*.

When *d = 0 to h*, **C$_p$** = -0.9

When *d = h to 2h*, **C$_p$** = -0.5

When *d ≥ h to 2h*, **C$_p$** = -0.3

When **C$_p$** is negative, it means the wind pressure is acting away from the building surface.

Since **P = qGC$_p$**, and **q** and **G** are the same for the entire flat roof, **C$_p$** determines the **P** on the flat roof. When the P is shown graphically, it looks like diagram A.

C$_{pi}$ is a factor for evaluating internal pressures.

C_{pi} = ±0.18 for a *completely* enclosed building.

C_{pi} = ±0.55 for a *partially* enclosed building.

If a building is symmetric, the internal pressures on opposing walls will cancel out.

q is the wind stagnation factor.

$$q = 0.00256 K_z K_{zt} K_d v^2 I$$

K_z takes into account the combined effects of exposure, height, and wind gust. You can look it up in ASCE 7, table 6-3.

K_{zt} takes into account escarpments and hills close to the building.
$K_{zt} = 1$ for level ground.

K_d takes into account the type of structure being studied.
$K_d = 0.85$ for buildings

I is the importance factor.
You can look this up in *Minimum Design Loads for Buildings* published by ASCE.

Note:
A building is to be classified as one of the following:

open building
- $A_o \geq 0.8 A_g$ *for each wall.* A_o *is the total area of the openings in the wall, and* A_g *is the total gross area of the wall.*

partially enclosed building
- $A_o > 1.10 A_{oi}$, *and*

- $A_o > minimum\ 4\ sf,\ or\ > 0.01 A_g$, *whichever is smaller*

- $A_{oi}/A_{gi} \leq 0.20$

  A_o *and* A_g *have been defined earlier.*

  A_{oi} *is the total area of the openings in the building envelope (walls and roof), NOT including* A_o.

  A_{gi} *is the total gross area of the building envelope (walls and roof), NOT including* A_g.

enclosed building
- *This is a building that is neither open nor partially enclosed.*

69. Answer: a

With regard to the 6-story building in figure 3.24, location 1 or the base of the building has the greatest shear during an earthquake. The shear decreases as the building height increases, and reaches zero at the top of the building, or location 7.

70. Answer: b and d

The following statements are correct:

- The Richter Scale is more accurate than the Modified Mercalli Scale.
  The **Richter Scale** measures the *amount of energy* release or magnitude, and does not indicate damages to buildings; the **Modified Mercalli Scale** measures the *effect* of earthquakes on people and buildings or the intensity of earthquakes, and is *imprecise* and subjective.
- In the Richter Scale, an earthquake of magnitude 8 releases about 32 times more energy than one of magnitude 7.
  The Richter Scale is a base-10 logarithmic scale. Each number represents about 32 times the *energy* of the preceding number, and has a *shaking amplitude* 10 times larger than the preceding number.

The following statements are incorrect:

- The Modified Mercalli Scale is more accurate than the Richter Scale.
- In the Richter Scale, an earthquake of magnitude 8 releases about 10 times more energy than one of magnitude 7.
- In the Modified Mercalli Scale, an earthquake of magnitude 8 releases about 10 times more energy than one of magnitude 7.
- In the Modified Mercalli Scale, an earthquake of magnitude 8 releases about 32 times more energy than one of magnitude 7.

71. Answer: a, b, and c

The Seismic Design Category (SDC) of a structure is determined by its

- design spectral response acceleration coefficients
- occupancy category
- height

See IBC, 1613.5.6, Determination of seismic design category:

"Occupancy Category I, II or III structures located where the mapped spectral response acceleration parameter at 1-second period, S_1, is greater than or equal to 0.75 shall be assigned to Seismic Design Category E. Occupancy Category IV structures located where the mapped spectral response acceleration parameter at 1-second period, S_1, is greater than or equal to 0.75 shall be assigned to Seismic Design Category F. All other structures shall be assigned to a seismic design category based on their **occupancy category** and the **design spectral response acceleration coefficients**, S_{DS} and S_{D1}, determined in accordance with Section 1613.5.4 or the site-specific procedures of ASCE 7. Each building and structure shall be assigned to the more severe seismic design category in accordance with Table 1613.5.6(1) or 1613.5.6(2), irrespective of the fundamental period of vibration of the structure, T."

See the following link:
http://publicecodes.cyberregs.com/icod/ibc/2006f2/icod_ibc_2006f2_16_par125.htm?bu2
=undefined

The structural **height** also affects the S$_{DC}$. For example, some low-rise structures are exempt from earthquake requirements. Detached one and two family wood-framed dwellings not more than two stories in height need to be designed using IRC requirements instead of IBC requirements.

The following are incorrect answers:
- weight
- footprint

72. Answer: b and d
The following statements regarding the Seismic Design Category (SDC) of a structure are true:
- Buildings in SDC C need to meet more restrictive earthquake requirements than buildings in SDC A.
- SDC includes categories A through E.

Occupancy group I are buildings with a *low hazard* to human life if they fail, such as agricultural facilities, minor storage facilities, and temporary facilities.

Occupancy group II are other buildings *not* included in occupancy groups I, III, and IV.

Occupancy group III are buildings with a *substantial hazard* to human life if they fail, such as public assemblies with an occupant load over 300; elementary or secondary schools, or daycare facilities with an occupant load over 250; colleges or adult education facilities with an occupant load over 500; jail and detention facilities; power-generating, potable water-treatment facilities, waste water-treatment facilities, or other public facilities not included in occupancy group IV; buildings with sufficient quantities of explosives or toxic substances and not included in occupancy group IV; or health care facilities with 50 or more resident patients, but without surgery or emergency treatment facilities.

Occupancy group IV are buildings designated as *essential facilities*, such as hospitals or health care facilities with surgery or emergency treatment facilities; police, fire, and rescue stations or emergency vehicle garages; designated earthquake, hurricane, or other emergency shelters; designated communication, emergency preparedness, and operation centers for emergency responses; power-generating or other public facilities required as back up facilities for occupancy group IV structures; buildings with highly toxic substances exceeding the maximum allowable quantities of Table 307.1.(2); air traffic control centers, aviation control towers, or emergency aircraft hangers; structures with significant national defense functions; or water-treatment facilities required to maintain water pressure for fire suppression.

Seismic Design Category A are buildings of **all** occupancy groups in areas expecting minor ground shaking. (Good soil)

Seismic Design Category B are buildings of occupancy groups I, II, and III in areas expecting moderate ground shaking. (Stratified soil with good and poor soil)

Seismic Design Category C are buildings of occupancy groups IV in areas expecting moderate ground shaking, and buildings of occupancy categories I, II, and, III in areas expecting SEVERE ground shaking.

Seismic Design Category D are buildings of **all** occupancy groups in areas expecting severe and destructive ground shaking, but NOT located close to a major fault. (Sites with poor soil are a good example.)

Seismic Design Category E are buildings of occupancy groups I, II, and III in areas near major active faults. (Soil or rock is of no consequence.)

Seismic Design Category F are buildings of Occupancy Groups IV in areas near major active faults. (Soil or rock is of no consequence.)

The following are incorrect answers:
- Buildings in SDC A need to meet more restrictive earthquake requirements than Buildings in SDC C.
- SDC includes categories A through D.

73. Answer: b and c
Please pay attention to the word "NOT."

The following regarding an earthquake's impact on a structure, are NOT true, and therefore the correct answers:
- In general, a number of cycles of moderate acceleration, sustained over time, are much easier for a building to withstand than a single much larger peak.
- Low–frequency waves (higher than 10 hertz) tend to have high amplitudes of acceleration but small amplitudes of displacement, compared to high-frequency waves, which have small accelerations and relatively large velocities and displacements.

The following are true statements, but the incorrect answers for this particular question:
- In general, a number of cycles of moderate acceleration, sustained over time, can be much more difficult for a building to withstand than a single much larger peak.
- High–frequency waves (higher than 10 hertz) tend to have high amplitudes of acceleration but small amplitudes of displacement, compared to low-frequency waves, which have small accelerations and relatively large velocities and displacements.

See Chapter 4 of the **FEMA publication number 454 (FEMA454),** *Designing for Earthquakes: A Manual for Architects,* for FREE at the following link:
http://www.fema.gov/library/viewRecord.do?id=2418

74. Answer: b
The **bracketed duration** is defined as the time between the first and last peaks of motion that exceeds a threshold value, commonly taken as 0.05g.

See Chapter 4 of the **FEMA publication number 454 (FEMA454),** *Designing for Earthquakes: A Manual for Architects.*

75. Answer: d
Height is the most important determinant of a building's period.

"A rule of thumb is that the building period equals the number of stories divided by 10; therefore, period is primarily a function of building height."

See Chapter 4 of the **FEMA publication number 454 (FEMA454),** *Designing for Earthquakes: A Manual for Architects.*

The following will affect a building's period, but they are not as important as the building height:
• construction materials
• contents
• geometric proportions
• structural system
• weight

76. Answer: b
The natural period of ground varies from about 0.4 seconds to 2 seconds, depending on the nature of the ground.

See Chapter 4 of the **FEMA publication number 454 (FEMA454),** *Designing for Earthquakes: A Manual for Architects.*

77. Answer: c
When a vibrating or swinging object is given further pushes that are also at its natural period, its vibrations increase dramatically in response to even rather small pushes. This phenomenon is called **resonance**.

The following are incorrect answers:
• **Acceleration** is the rate at which the velocity of a body changes with time.
• **Inertia** is the tendency of an object to resist any change in its motion.
• **Velocity increase** means increase in speed.

78. Answer: a and d
The following statements regarding the earthquake's impact on a structure are true:
• Accelerations created by ground motion increase rapidly as the building damping value decreases.
• Damping is measured by reference to a theoretical damping level termed **critical damping**.

The following are incorrect answers:

- Accelerations created by ground motion decrease rapidly as the building damping value decreases.
- Damping is measured by reference to a theoretical acceleration level termed critical acceleration.

See Chapter 4 of the **FEMA publication number 454 (FEMA454)**, *Designing for Earthquakes: A Manual for Architects*.

79. Answer: b and c

The following statements regarding earthquakes are true:

- P wave is a primary wave that alternately pushes (compresses) and pulls (dilates) the rock.
- P waves, just like acoustic waves, are able to travel through solid rock, such as granite and alluvium; through soils; and through liquids, such as volcanic magma or the water of lakes and oceans.

The following are incorrect answers:

- P wave is a wave that moves perpendicular to the ground surface.
- S waves, just like acoustic waves, are able to travel through solid rock, such as granite and alluvium; through soils; and through liquids, such as volcanic magma or the water of lakes and oceans.

Per Chapter 2, Section 2.3.2, Types of Earthquake Waves of **FEMA publication number 454 (FEMA454)**, *Designing for Earthquakes: A Manual for Architects*:

"The first two types of waves travel through the body of the earth before arriving at the surface. The faster of these "body" waves is appropriately called the **primary** or **P wave**. Its motion is the same as that of a sound wave in that, as it spreads out, it alternately pushes (compresses) and pulls (dilates) the rock. These P waves, just like acoustic waves, are able to travel through solid rock, such as granite and alluvium, through soils, and through liquids, such as volcanic magma or the water of lakes and oceans.

The second and slower seismic body wave through the earth is called the **secondary** or **S wave** or sometimes the **shear wave**. As an S wave propagates, it shears the rocks sideways at right angles to the direction of travel. At the ground surface, the upward emerging S waves also produce both vertical and horizontal motions. Because they depend on elastic shear resistance, S waves cannot propagate in liquid parts of the earth, such as lakes. As expected from this property, their size is significantly weakened in partially liquefied soil. The speed of both P and S seismic waves depends on the density and elastic properties of the rocks and soil through which they pass. In earthquakes, P waves move faster than S waves and are felt first. The effect is similar to a sonic boom that bumps and rattles windows. Some seconds later, S waves arrive with their significant component of side-to-side shearing motion...for upward wave incidence, the ground shaking in the S waves becomes both vertical and horizontal, which is the reason that the S wave motion is so effective in damaging structures."

80. Answer: b and c

The following terms are used to define the earthquake wave:
- Love wave
- Rayleigh wave

The following are incorrect answers:
- H wave
- W wave

Per Chapter 2, Section 2.3.2, Types of Earthquake Waves of **FEMA publication number 454 (FEMA454),** *Designing for Earthquakes: A Manual for Architects*:

"The first type of surface wave is called a **Love wave**...Its motion is the same as that of S waves that have no vertical displacement; it moves the ground side to side in a horizontal plane parallel to the earth's surface, but at right angles to the direction of propagation. The second type of surface wave is called a **Rayleigh wave**...Like ocean waves, the particles of rock displaced by a Rayleigh wave move both vertically and horizontally in a vertical plane oriented in the direction in which the waves are traveling. The motions are usually in a retrograde sense...Each point in the rock moves in an ellipse as the wave passes."

81. Answer: a and c

At a field visit, an architect notices that the contractor has completely filled the space between two structural columns with a rigid wall per the owner's instruction. The architect can do the following:
- Advise the owner to have the structural engineer review the changes and submit revised plans to the city.
- Advise the owner that this may have created a structural problem and order the contractor to rip out the wall after the client agrees this is an unnecessary element.

The following are incorrect answers:
- Do nothing because this is a very minor revision, has no impact on the structural integrity of the building, and does not even require a building permit.
- Revise the architectural plans and submit the architectural plans to the city since the revision does not involve structural plans.

By filling the space between two structural columns with a rigid wall, the owner has inadvertently created a **short-column condition**.

Per Chapter 4 of **FEMA publication number 454 (FEMA454),** *Designing for Earthquakes: A Manual for Architects*:

"Such a simple act of remodeling may not seem to require engineering analysis, and a contractor may be hired to do the work: often such work is not subject to building department reviews and inspection. Serious damage has occurred to buildings in earthquakes because of this oversight."

Suppose two columns have the same cross section, but one column is half the length of the other. Mathematically, the stiffness of a column varies approximately as the cube of its length. Therefore, **the short column will be eight times stiffer (2^3) instead of twice as stiff** and will be subject to eight times the horizontal load of the long column. Stress is concentrated in the short column, while the long column is subject to nominal forces.

82. Answer: c

Per Chapter 7, Section 7.7.5, Configurations Are Critical, of **FEMA publication number 454 (FEMA454)**, *Designing for Earthquakes: A Manual for Architects*:

"Configuration, or the three-dimensional form of a building, frequently is the governing factor in the ultimate seismic behavior of a particular structure."

The following are incorrect answers:
- the weight of the structure
- the height of the structure (This answer is somewhat correct, but NOT complete, and therefore not the best answer.)
- none of the above

83. Answer: a

The following is the correct order to arrange structural systems from low post-earthquake repair cost to high post-earthquake repair cost:
- seismic isolation, dampers plus steel moment-resisting frame, un-bonded steel brace, timber framing, ductile steel moment-resisting frame, steel frame plus braces

The following are incorrect answers:
- seismic isolation, dampers plus steel moment-resisting frame, steel frame plus braces, un-bonded steel brace, timber framing, ductile steel moment-resisting frame
- seismic isolation, dampers plus steel moment-resisting frame, steel frame plus braces, ductile steel moment-resisting frame, un-bonded steel brace, timber framing
- seismic isolation, dampers plus steel moment-resisting frame, ductile steel moment-resisting frame, un-bonded steel brace, timber framing, steel frame plus braces

See Chapter 7, figure 7-11A, Structural seismic characteristics, and figure 7-11B, Structural seismic characteristics, of **FEMA publication number 454 (FEMA454),** *Designing for Earthquakes: A Manual for Architects*

84. Answer: The three basic alternative types of vertical lateral force–resisting systems are **shear walls, braced frames,** and **moment-resistant frames.**

There are two general types of **braced frames:** conventional concentric and eccentric. In the **concentric frame**, the centerlines of the bracing members meet the horizontal beam at a single point. In the **eccentric braced frame**, the braces are deliberately designed to meet the beam some distance apart from one another. The short piece of beam between the ends of the braces is called a link beam. The purpose of the link beam is to provide ductility to the system. Under heavy seismic forces, the link beam will distort and dissipate the energy of the earthquake in a controlled way, thus protecting the remainder of the structure.

See Chapter 5, 5.2.1, The Vertical Lateral Resistance Systems, of **FEMA publication number 454 (FEMA454),** *Designing for Earthquakes: A Manual for Architects.*

85. Answer: c

The term used to identify a horizontal-resistance member that transfers lateral forces between vertical-resistance elements (shear walls or frames) is **diaphragm.**

A diaphragm that forms part of a resistant system may act either in a flexible or rigid manner, depending partly on its size (the area between enclosing resistance elements or stiffening beams) and also on its material. The flexibility of the diaphragm, relative to the shear walls whose forces it is transmitting, also has a major influence on the nature and magnitude of those forces. With **flexible diaphragms** made of wood or steel decking without concrete, walls *take loads according to tributary areas* (if mass is evenly distributed). With **rigid diaphragms** (usually concrete slabs), *walls share the loads in proportion to their stiffness.*

The following are incorrect answers:
- beam
- space frame
- floor system

See Chapter 5, 5.2.2, Diaphragms—the Horizontal Resistance System, of **FEMA publication number 454 (FEMA454),** *Designing for Earthquakes: A Manual for Architects.*

86. Answer: b

Indeed many engineers believe that it is the architectural irregularities that contribute primarily to poor seismic performance and occasional failure.

The following are incorrect answers:
- regular building shapes
- weights of the buildings
- heights of the buildings

See Chapter 5, 5.2.3, Optimizing the Structural/Architectural Configuration, of **FEMA publication number 454 (FEMA454),** *Designing for Earthquakes: A Manual for Architects.*

87. Answer: e

All of the following contribute to poor seismic performance and occasional failure:
- re-entrant corners
- soft or weak floors
- torsion

- short-column phenomenon

See Chapter 5, 5.3.1, Stress Concentrations, of **FEMA publication number 454 (FEMA454),** *Designing for Earthquakes: A Manual for Architects.*

88. Answer: e
Please pay attention to the world "NOT."

Circular plans do NOT have the potential to seriously impact seismic performance.

The following have the potential to seriously impact seismic performance, and are therefore the incorrect answers:
- soft and weak stories
- discontinuous shear walls
- variations in perimeter strength and stiffness
- re-entrant corners

89. Answer: a, b, and c
The International Style often has a number of characteristics not present in earlier frame and masonry buildings that has led to poor seismic performance:
- the elevation of the building on stilts or pilotis (Without full understanding of the seismic implications of vertical structural discontinuity, designers often create soft and weak stories.)
- the free plan and elimination of interior-load bearing walls (The replacement of masonry and tile partitions by frame and gypsum board greatly reduces the energy absorption capability of the building and increases its drift, leading to greater nonstructural damage and possible structural failure.)
- the great increase of exterior glazing and the invention of the light-weight curtain wall (Like free interior planning, the light exterior cladding greatly reduces the energy-absorption capability of the building and increases its drift.)

The following are incorrect answers:
- the building plan configuration (International Style buildings are frequently close to ideal seismic building configuration.)
- the elimination of building decorations and related redundancy of building components (Building decorations have nothing to do with redundancy in building structural components. This is an invented answer used as a distracter.)

90. Answer: c
The first seismic code was created in the United States in the 1920s.

In 1927, the Pacific Coast Building Officials Conference (precursor to ICBO, the International Conference of Building Officials) included an appendix of optional seismic design provisions in the first edition of the Uniform Building Code (UBC). A lateral load requirement was set at 7.5% of the building weight with an increase to 10% for sites with soft soils. This established the first version of the **equivalent lateral force procedure** still used in seismic codes today.

91. Answer: a, b, e, and f

Technical performance levels translate qualitative performance levels into damage states expected for structural and nonstructural systems. As defined in table 6-2, the Structural Engineers Association of California (SEAOC) Vision 2000 document proposes four qualitative performance levels:

- fully operational
- operational
- life safety
- near collapse

The following are incorrect answers:

- moderately damaged
- severely damaged

See Chapter 6, 6.5.2, Performance Levels, of **FEMA publication number 454 (FEMA454),** *Designing for Earthquakes: A Manual for Architects.*

92. Answer: b

"A philosophy quickly developed suggesting that existing buildings be treated differently from new buildings with regard to seismic requirements. First, archaic systems and materials would have to be recognized and incorporated into the expected seismic response, and secondly, due to cost and disruption, seismic design force levels could be smaller. The smaller force levels were rationalized as providing minimum life safety, but not the damage control of new buildings, a technically controversial and unproven concept, but popular. Commonly existing buildings were then designed to 75% of the values of new buildings—a factor that can still be found, either overtly or hidden, in many current codes and standards for existing buildings."

See Chapter 8, 8.2.2, Philosophy Developed for Treatment of Existing Buildings, of **FEMA publication number 454 (FEMA454),** *Designing for Earthquakes: A Manual for Architects.*

93. Answer: a, b, and c

As the conceptual framework of evaluation and retrofit developed, legal and code requirements were also created. These policies and regulations can be described in three categories:

- active
- passive
- post-earthquake

The following are incorrect answers:

- preventive
- retrofit

See Chapter 8, 8.2.3, Code Requirements Covering Existing Buildings, of **FEMA publication number 454 (FEMA454)**, *Designing for Earthquakes: A Manual for Architects.*

94. Answer: c and d

The following statements are true:

- After an earthquake, damaged buildings should be either Green-tagged or Yellow-tagged. (Some slightly damaged buildings are Green-tagged.)
- After an earthquake, a damaged building that creates a public risk and requires immediate mitigation should be Red-tagged.

The following are incorrect answers:

- After an earthquake, all buildings should be inspected by a building official. (Some undamaged buildings are NOT inspected at all because they are NOT required to be by codes.)
- After an earthquake, all buildings should be inspected and Green-tagged by a building official if they are undamaged. (See previous explanation.)

95. Answer: a and c

The followings statements are true:

- The principal seismic risk in the United States comes from the existing building stock.
- The FEMA "yellow book" series are less known to architects.

The following are incorrect answers:

- The principal seismic risk in the United States comes from selecting the wrong occupancy category when designing buildings.
- The FEMA "yellow book" series are less known to engineers.

See Chapter 8, 8.3, The FEMA Program to Reduce the Seismic Risk from Existing Buildings, of **FEMA publication number 454 (FEMA454)**, *Designing for Earthquakes: A Manual for Architects.*

96. Answer: b

Considering only the benefit-cost ratio of seismic retrofit, tilt-up buildings are most likely to be retrofitted.

The following are incorrect answers:

- unreinforced masonry (URM) buildings
- wood-stud buildings
- moment-resisting frame buildings

Unreinforced masonry (URM) buildings were a popular building type early in the twentieth century and are now recognized as perhaps the worst seismic performer as a class. However, they were not outlawed in the zones of high seismicity until the 1933 code, and continued to be built in much of the country with no significant seismic design provisions until quite recently. Unreinforced masonry (URM) buildings are very costly for rehabilitation.

Rehabilitation of a tilt-up building is usually fairly inexpensive, and usually proves to be cost effective.

Wood-stud buildings and moment-resisting frame buildings are good seismic performers and are unlikely candidates for seismic retrofit.

97. Answer: a and b
The following are flexible diaphrams:
- metal deck roof over roof joists over wood trusses
- Plywood over floor joists

The following are incorrect answers:
- cast-in-place concrete floor
- pre-cast concrete floor

98. Answer: a
According to FEMA 454, the expected lifespan of a building is 50 years.

See Chapter 8, 8.4.3, Other Evaluation Issues, paragraph entitled "Describing Shaking Intensity" of **FEMA publication number 454 (FEMA454),** *Designing for Earthquakes: A Manual for Architects.*

99. Answer: d
According to FEMA 454, nationally applicable building codes are based on the level of shaking intensity expected at any site once every 500 years (on average).

See Chapter 8, 8.4.3, Other Evaluation Issues, paragraph entitled "Describing Shaking Intensity" of **FEMA publication number 454 (FEMA454),** *Designing for Earthquakes: A Manual for Architects.*
You should look through the figures and photos of the remaining chapters of FEMA454, This will help you in both the SS ARE exam prep and in real practice.

100.   Answer: a
Please pay attention to the word "not." We are looking for the statement that is not true.

The following statement is not true, and therefore the correct answer:
- The seismic performance of nonstructural systems is not important since it will not affect life safety.

The following statements are true, but they are the incorrect answers:
- The seismic performance of nonstructural systems is important since it will affect property loss and/or life safety.

- Historically, the seismic performance of nonstructural systems and components has received little attention from designers.
- Some investigators have postulated that nonstructural system or component failure may lead to more injury and death in the future than structural failure.

See Chapter 9, 9.2.3 Consequences of Inadequate Nonstructural Design of **FEMA publication number 454 (FEMA454),** *Designing for Earthquakes: A Manual for Architects:*

"Historically, the seismic performance of nonstructural systems and components has received little attention from designers. The 1971 San Fernando earthquake alerted designers to the issue' mainly because well-designed building structures were able to survive damaging earthquakes while nonstructural components suffered severe damage. It became obvious that much more attention had to be paid to the design of nonstructural components. Some investigators have postulated that nonstructural system or component failure may lead to more injury and death in the future than structural failure."

101. Answer: a
The architect is responsible for most of the nonstructural seismic design issues for both systems and components.

The following are incorrect answers:
- the structural engineer
- the electrical engineer
- the mechanical engineer
- the specialty consultant

See Chapter 9, 9.8, WHO is responsible for DESIGN? Table 9-3, Design Responsibilities for Nonstructural Components of **FEMA publication number 454 (FEMA454),** *Designing for Earthquakes: A Manual for Architects.*

102. Answer: d
For a single family residence, all of the following should be connect to UFER or other kinds of ground:
- electrical lines
- cable lines
- phone lines

103. Answer: d
Please pay attention to the word "EXCEPT." We are looking for the incorrect procedure.

The permitted wind design procedure includes all of the following EXCEPT:
- diagram procedure (This is an invented term and used as a distracter.)

The following are incorrect answers:
- simplified procedure

- analytical procedure
- wind tunnel procedure

See page 25 of the PDF file of "Wind Design Made Simple" by ICC TRI-Chapter Uniform Code Committee available for FREE at the following link: http://www.calbo.org/Documents/SimplifiedWindHandout.pdf

104. Answer: d
Inland waterways belong to wind exposure category D per ASCE 7.

ASCE 7 is a book entitled *Minimum Design Loads for Buildings and Other Structures*, published by ASCE (American Society of Civil Engineers). You can purchase it from Amazon, but I doubt you'll refer to it a lot in practice, because you normally leave the detailed structural calculations to your structural engineer.

Per ASCE 7, 1609.4 & 6.5.6 Exposure Categories are defined as follows.
- **Exposure A** is no longer used in ASCE 7.
- **Exposure B** is used as the default.
- **Exposure C** includes shorelines of hurricane prone regions (no longer Exposure D).
- **Exposure D** now only applies to inland waterways, Great Lakes, Coastal California, Oregon, Washington, and Alaska.

See page 20 of the PDF file of "Wind Design Made Simple" by ICC TRI-Chapter Uniform Code Committee available for FREE at the following link: http://www.calbo.org/Documents/SimplifiedWindHandout.pdf

IBC also has similar definitions.
"An exposure category shall be determined in accordance with the following:

Exposure B. Exposure B shall apply where the ground surface roughness condition, as defined by Surface Roughness B, prevails in the upwind direction for a distance of at least 2,600 feet (792 m) or 20 times the height of the building, whichever is greater.

Exception: For buildings whose mean roof height is less than or equal to 30 feet (9144 mm), the upwind distance is permitted to be reduced to 1,500 feet (457 m).

Exposure C. Exposure C shall apply for all cases where Exposures B or D do not apply.

Exposure D. Exposure D shall apply where the ground surface roughness, as defined by Surface Roughness D, prevails in the upwind direction for a distance of at least 5,000 feet (1524 m) or 20 times the height of the building, whichever is greater. Exposure D shall extend inland from the shoreline for a distance of 600 feet (183 m) or 20 times the height of the building, whichever is greater."

See related IBC information for FREE at the following link: http://publiccodes.cyberregs.com/icod/ibc/2006f2/icod_ibc_2006f2_16_par100.htm

105.   Answer: c
Figure 3.25 shows a reinforced concrete wall. The following statement is correct:
• Pier 1 and Pier 4 resist more lateral load than Pier 2 and Pier 3.

The following are incorrect answers:
• Pier 1 and Pier 2 resist more lateral load than Pier 3 and Pier 4.
• Pier 1 and Pier 3 resist more lateral load than Pier 2 and Pier 4.
• Pier 3 and Pier 4 resist more lateral load than Pier 1 and Pier 2.
• Each pier resists the same lateral load.

The lateral load will be distributed to the piers according to their rigidities. Rigidity is defined as resistance to deflection. For a reinforced concrete wall, the rigidity is determined by the width to height ratio. Since all piers have the same height (4 feet), the wider pier will have a greater rigidity. Pier 1 and Pier 4 are both 8 feet wide, and Pier 3 and Pier 4 are both 6 feet wide, so, Pier 1 and Pier 4 resist more lateral load than Pier 2 and Pier 3.

106.   Answer: c
Figure 3.26 shows a 2-story room addition to a 1-story existing building. The following statement is correct:
• Column C will resist more lateral load than the other columns.

The following are incorrect answers:
• Column A will resist more lateral load than the other columns.
• Column B will resist more lateral load than the other columns.
• Column D will resist more lateral load than the other columns.
• Each column resists the same lateral load.

This room addition project accidentally created a "short column" condition at column C. Because column C is shorter and more rigid than the other columns, it will resist more lateral load and has a greater risk of failure.

The weakest location is at the connection of columns B and C.

107.   Answer: b
1/4" per foot (1%) is a minimum slope for a flat roof to avoid ponding, the accumulation of excessive water causing excessive deflection and potential failure of the roof.

108.   Answer: b
The wind maps referred to by IBC and ASCE are based on a yearly 2% probability of occurrence.

Per IBC, Section1609.3:
"In non-hurricane-prone regions, when the basic wind speed is estimated from regional climatic data, the basic wind speed shall be not less than the wind speed associated with an annual probability of 0.02 (50-year mean recurrence interval), and the estimate shall be

adjusted for equivalence to a 3-second gust wind speed at 33 feet (10 m) above ground in Exposure Category C. The data analysis shall be performed in accordance with Section 6.5.4.2 of ASCE 7."

See IBC, Section1609.3, Basic wind speed, for FREE at the following link:
http://publicecodes.cyberregs.com/icod/ibc/2006f2/icod_ibc_2006f2_16_par095.htm

109.   Answer: a
Diaphragm boundary is a term frequently used in light-frame construction.

See IBC, 1602.1 Definitions, for FREE at the following link:
http://publicecodes.cyberregs.com/icod/ibc/2006f2/icod_ibc_2006f2_16_sec002.htm

Based on feedback from other candidates, you have a good chance of getting quite a few questions regarding these definitions on the real exam. You really need to become very familiar with them.

The following are incorrect answers:
- heavy timber construction
- masonry construction
- tilt-up construction

110.   Answer: b
Per IBC, "where the live loads for which each floor or portion thereof of a commercial or industrial building is or has been designed to exceed 50 psf (2.40 kN/m^2), such design live loads shall be conspicuously posted by the owner in that part of each story in which they apply, using durable signs. It shall be unlawful to remove or deface such notices."

See IBC, 1603.3 Live loads posted, for FREE at the following link:
http://publicecodes.cyberregs.com/icod/ibc/2006f2/icod_ibc_2006f2_16_par013.htm

111.   Answer: a
In the United States, wind causes the most damage to buildings. In fact, annually, wind damages to buildings/structures exceed all other natural disasters combined.
The following are incorrect answers:
- earthquake
- flood
- mountain fire

112.   Answer: d
During a field visit, a building inspector asks the contractor to nail the roof sheathing to the blocking along the exterior wall. This is to prevent potential damages caused by both wind and earthquake.

113.   Answer: c
       According to basic wind speeds, a 3-second gust is <u>90</u> mph for almost all of the continental United States, except the east and west coast.

114.   Answer: c
       Figure 3.27 shows a 3-story moment-resisting frame with hinged bases resisting lateral loads. Ignoring the dead load, tension in one hinge base and compression in another hinge base resist the overturning caused by the lateral loads (figure 4.11).

       The following are incorrect answers:
       • shear in the columns (Shear in the columns does NOT resist the overturning.)
       • moment at the column bases (There is no moment at any hinged bases.)
       • tension in the beams (The beams have no tension since we ignore the dead loads.)
       • compression in the beams (Compression in the beams does NOT resist the overturning.)

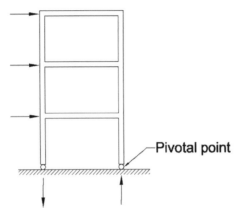

Figure 4.11   A moment-resisting frame with hinged bases resisting lateral loads

115.   Answer: f and h
       Referring to the same 3-story moment-resisting frame as shown in figure 3.27, if we include the dead load, the following resist the overturning caused by the lateral loads:
       • weight of a column and beams (This answer was placed at the end to make sure you were patient enough to read all the choices. See figure 4.12.)
       • tension or compression in one hinge base, and compression in another hinge base (Because we include the dead loads, one of the hinge bases has compression, but the other hinge base may have tension OR compression, depending on the actual amount of the dead load.)

       The following are incorrect answers:
       • shear in the columns
       • weight of the columns (This answer is not complete. It does not include the weight of the beams.)
       • moment at the column bases (There is no moment at hinge base.)
       • tension in one hinge base, and compression in another hinge base
       • tension in the beams

- compression in the beams

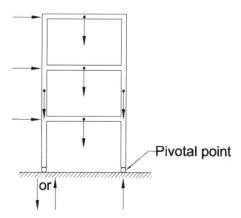

Figure 4.12   A moment-resisting frame with hinged bases and dead loads resisting lateral loads

116.   Answer: d
The rigid frame shown has moment-resisting connections and fixed bases, so both the upper corners and the two fixed bases are restrained against rotation, as shown in the correct answer, Diagram D.

If the column bases were hinged, the bases would provide no restraint against rotation, and Diagram B would be the correct answer.

117.   Answer: c
The rigid frame shown has moment-resisting connections and fixed bases, so both the upper corners and the two fixed bases are restrained against rotation, as shown in the correct answer, Diagram C.

If the column bases were hinged, the bases would provide no restraint against rotation, and Diagram A would be the correct answer.

If the column bases were hinged *and* the bases could slide, the bases would provide no restraint against rotation or horizontal reactions, making Diagram D the correct answer.

Diagram B is a simple beam supported by columns without rigid joints.

118.   Answer: a
A horizontal force of 25 kips is applied to a frame with diagonal braces made of cables as shown in figure 3.30. The internal tension in each brace is as follows:
- brace 1 = 0, brace 2 = 35.36 kips.

The key to this question is that diagonal braces made of cables can *only* resist tension, but not compression. So, Brace 1 will always have zero stress since it cannot resist compression.

See the free body diagram as shown in figure 4.13.

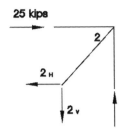

Fig 4.13

Figure 4.13   Free body diagram of a frame

$\sum H = 0$

$+ 25\text{kips} - 2_H = 0$

$2_H = 25$ kips

Since the frame is 16' wide and 16' high, the brace is at an 45° angel, so $2_v = 25$ kips

Therefore, the tension in brace 1 = $\sqrt{25^2 + 25^2} = 35.36$ kips

119.   Answer: a and d
Figure 3.31 shows a moment-resisting frame. The following statements are correct:
- The bottom of the column is free to rotate. (There are only two anchor bolts, and they offer little resistance to rotation.)
- The top of the column is fixed against rotation. (This is a moment frame system.)

The top of the column can still translate, or move horizontally.

The following are incorrect answers:
- The top of the column is free to rotate.
- The bottom of the column is fixed against rotation.

120.   Answer: d
Diagram D in figure 3.32 shows the distribution of lateral forces used in seismic design. The diagram is normally parabolic. When all stories are the same and the building is short and stiff, the diagram is a reversed triangle, with the maximum lateral force at the top and zero force at the base.

121.   Answer: b and d
Systems B and D shown in figure 3.33 are stable under lateral forces.

System A will collapse under vertical or lateral forces or the dead load of the members. System C will collapse under lateral forces.

122.   Answer: c
Figure 3.34 shows a building's floor plan. The total wind load in the east-west direction is smaller than in the north-south direction.

The total east and west walls are 60' long, and the north and south walls are 120' long. So, the surface area subject to east-west wind is about half of the surface area subject to north-south wind. Therefore, the total wind load in the east-west direction is smaller than in the north-south direction.

The gust factor is subject to building height and exposure, and not subject to the wind direction. So, answer "d" is incorrect.

123.   Answer: a
Figure 3.35 shows beam to column connections. The connection as shown in Diagram I is part of a moment-resisting frame.

To be part of a moment-resisting frame, the beam flange *must* be welded directly to the columns or rigidly attached by plates welded to the columns and bolted to the beam.

The connection, as shown in Diagram II shows a beam seat transferring only the shear from the beam webs to the columns. The top angle is only used to hold the top of the beam in place, and is not a rigid connection to the column.

124.   Answer: c
Figure 3.36 shows a truss. Diagram C shows the correct reactions at A and B.

First of all, support B is on a roller, so the horizontal reaction at support B has to be zero, and the horizontal reaction at support A has to equal 500lb to keep the truss in balance. So, only A and C are possible answers.

Take the moment about support A:
$(500 \text{ lb} \times 8 \text{ ft}) + (V_B \times 32 \text{ ft}) = (500 \text{ lb} \times 32 \text{ ft})$

Note: V_B is the vertical reaction at support B.

$V_B (32 \text{ ft}) = (500 \text{ lb} \times (32 \text{ ft}) - (500 \text{ lb} \times 8 \text{ ft}) = 12,000 \text{ lb-ft}$
$V_B = 12,000 \text{ lb-ft}/32 \text{ ft} = 375 \text{ lb}$
So, the vertical reaction at support B is 375 lb.

$V_A + V_B = 500 \text{ lb}$

Note: V_A is the vertical reaction at support A.

$V_A + 375 \text{ lb} = 500 \text{ lb}$
$V_A = 500 \text{ lb} - 375 \text{ lb} = 125 \text{ lb}$

So, the vertical reaction at support A is 125 lb.
Therefore, Diagram C is the correct answer.

125.   Answer: d
        The net uplift at point 1 is zero.

        We take moment at point 2:
        Overturning moment = 500 lb x 10 ft = 5,000 ft-lb

        Dead load resisting moment = 500 lb x 20 ft = 10,000 ft-lb > Overturning moment

        Therefore, there is zero uplift at Point 1.

B. Mock Exam Solution: SS Graphic Vignette Section

    1. Step-by-step description for a passing solution using the official NCARB SS practice program

1) Let us start with the lower level. Click on **layers** to make sure we set the current layer to the lower level (figure 4.14).

   Note:

   Once you draw an element on a level, it is impossible to move it to another level. So, this step is VERY important. If you draw elements on the wrong level, you may waste a lot of time and not have enough time to redraw them on the correct level, causing you to fail the vignette and the SS ARE division.

2) Click on **cursor** to set the cursor to full-screen mode (figure 4.15). This will make it easier for you to align structural elements.

3) Use **Draw > Bearing Wall w/ Bond Beam** to draw the bearing walls (figure 4.16). Make sure you do not accidentally cover the openings. Use **Draw > Column** to place one column on the east wall of the Common Area.

   Note:
   - *Pay attention to the column we draw. We need it to support the lintel for the small door.*
   - *Use **zoom** to zoom in, and use **move, adjust** to change the bearing wall if necessary.*

4) Use **Draw > Beam or Lintel** to place lintels over all openings in the bearing walls (figure 4.17). Make sure that each lintel is supported by bearing walls or columns on both ends.

   Note:

   Pay attention to the short lintel for the small door on the east wall of the Common Area. We need it to support the wall above.

5) The window wall and the clerestory window extend to the underside of the structure above. All other openings (including the opening between the Common Area and Covered Entry) have a head height of 7 ft above finished floor. Therefore, we need to add a beam or lintel for the opening between the Common Area and Covered Entry. Use **Draw > Beam or Lintel** to place a long lintel over this opening and span it between two bearing walls (figure 4.18). This beam will also support the upper wall between the Common Area and Covered Entry.

   Note:

   You can also make the wall between the Common Area and Covered Entry a bearing wall, and use a shorter lintel.

6) Use **Draw > Joists > Select joist direction > Spacing at 48" o.c.** to draw joists between the bearing walls (figure 4.19).

Note:

- *The joists are drawn as a 2-point rectangle.*
- *Make sure both ends of the joists are supported by bearing walls, beams, or lintels.*
- *Make sure the joists support the edge of the roof at locations with no bearing walls, beams, or lintels.*
- *Per the program, the metal roof deck is capable of carrying the design loads on spans up to and including 4 ft. Therefore, the maximum spacing between the joists is 4 ft or 48". We set the spacing at 48".*

7) Use **Draw > Decking > Select deck direction** to draw decks between the joists (figure 4.20).

Note:

- *The decks are drawn as a 2-point rectangle.*
- *Make sure that both ends of the decks are supported by joists.*
- *The grey-color double arrow indicates the direction of the deck. It should be perpendicular to the joists.*

8) The lower level is completed. Let us start with the upper level. Click on **layers** to open a dialogue box, set the **Current Level** to **Upper,** and then click **OK** (figure 4.21).

9) Use **Draw > Column** to draw six columns on top of the bearing walls. Align the columns (figure 4.22).

Note:

- *Once you set the **Current Level** to **Upper** and select the other level as visible, the elements on the lower level show in grey color.*
- *The column layout accommodates the clerestory window located along the full length of the north wall of the common area.*

10) Use **Draw > Beam or Lintel** to place beams between the columns along the direction of the bearing walls (figure 4.23).

Note:
Place beams from the center of a column to the center of an adjacent column.

11) Use **Draw > Joists > Select joist direction > Spacing at 48" o.c.** to draw joists between the bearing beams (figure 4.24).

12) Use **Draw > Decking > Select deck direction** to draw decks between the joists (figure 4.25).

13) This is your final solution (figure 4.26).

14) Of course, you can also solve this vignette using only columns and no bearing walls. Just make sure your upper level columns are supported by lower level columns at the same location (figure 4.27 & figure 4.28).

Note:
- *Pay attention to the beam on the west wall of the Common Area at the lower level. We need it to support the wall above.*
- *Pay attention to the short lintel for the small door on the east wall of the common area. We need it to support the wall above.*

2. Notes on graphic vignette traps

1) Make sure you draw the elements on the correct level.
2) Pay attention to the column that we need to support the lintel for the small door on the east wall of the Common Area.
3) Pay attention to the short lintel for the small door on the east wall of the Common Area.
4) Make sure that both ends of the joists are supported by bearing walls, beams, or lintels.
5) Make sure the joists support the edges of the roof at locations with no bearing walls, beams, or lintels.
6) Per the program, the metal roof deck is capable of carrying the design load on spans up to and including 4 ft. Therefore, the maximum spacing between the joists is 4 ft or 48" o.c. We set the spacing at 48".
7) Make sure both ends of the deck are supported by joists.
8) The grey-color double arrow indicates the direction of the deck. It should be perpendicular to the joists.

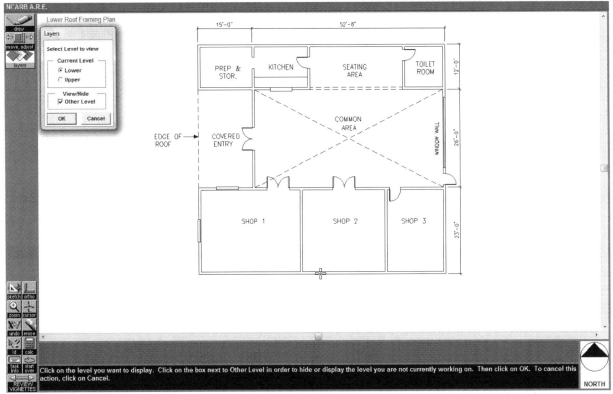

Figure 4.14   Click on **layers** to make sure we set the current layer to the lower level.

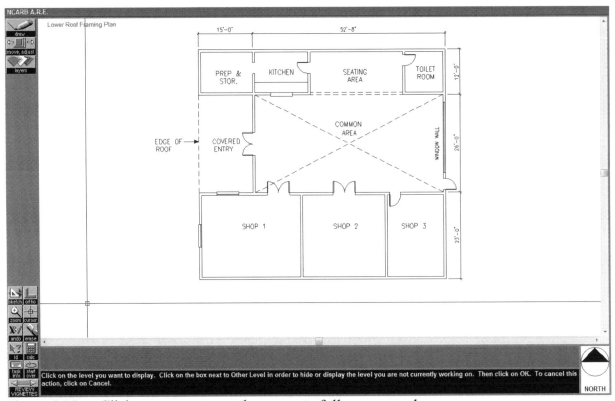

Figure 4.15   Click on **cursor** to set the cursor to full-screen mode.

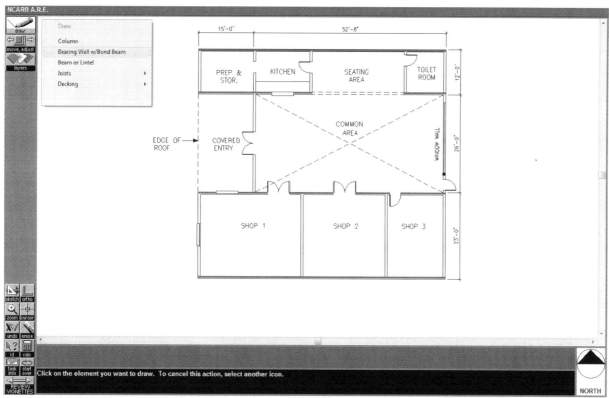

Figure 4.16   Use **Draw > Bearing Wall w/ Bond Beam** to draw the bearing walls.
Use **Draw > Column** to place one column on the east wall of the Common Area.

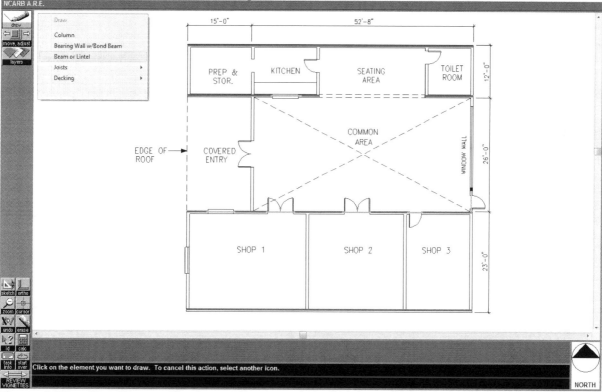

Figure 4.17   Use **Draw > Beam or Lintel** to place lintels over all openings in the bearing walls.

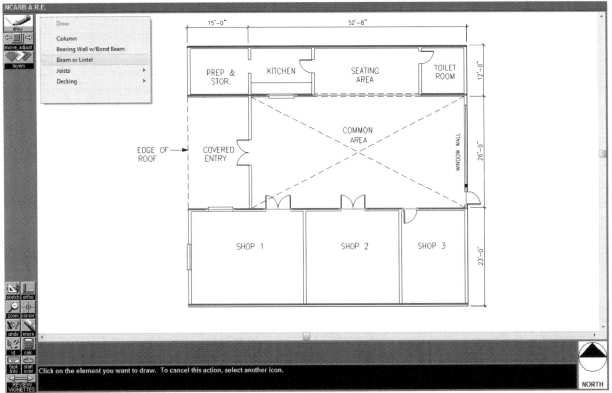

Figure 4.18   Use **Draw > Beam or Lintel** to place a long lintel over the opening between the Common Area and Covered Entry and span it between two bearing walls.

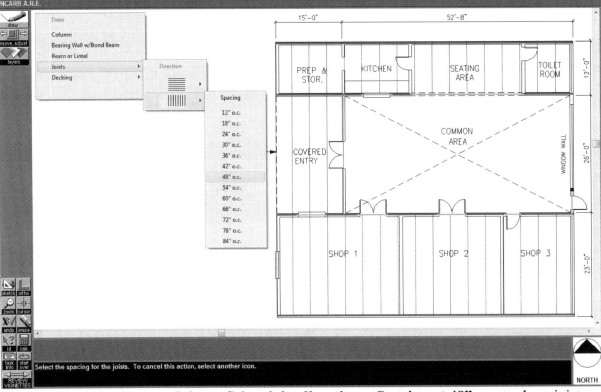

Figure 4.19   Use **Draw > Joists > Select joist direction > Spacing at 48" o.c.** to draw joists between the bearing walls.

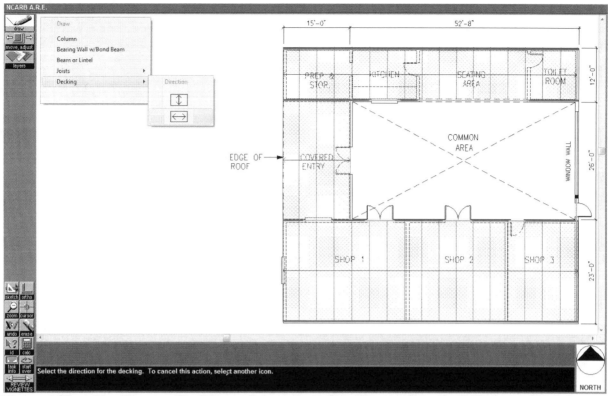

Figure 4.20   Use **Draw > Decking > Select deck direction** to draw decks between the joists.

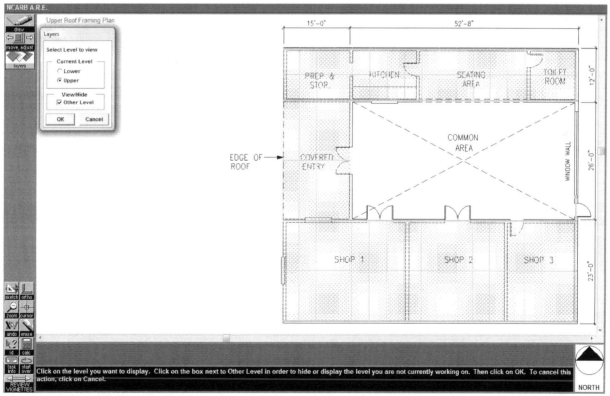

Figure 4.21   Click on **layers** to open a dialogue box, set the **Current Level** to **Upper,** and then click **OK.**

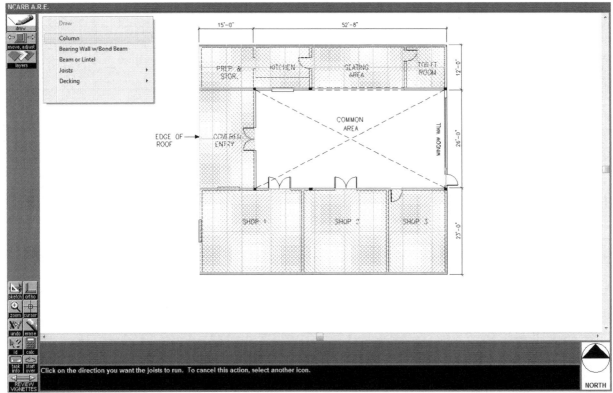

Figure 4.22   Use **Draw > Column** to draw six columns on top of the bearing walls. Align the columns.

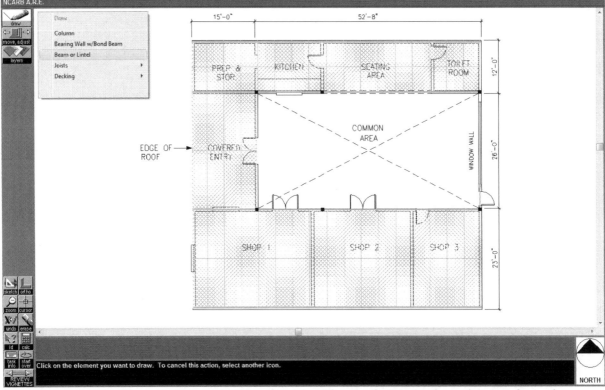

Figure 4.23   Use **Draw > Beam or Lintel** to place beams between the columns along the direction of the bearing walls.

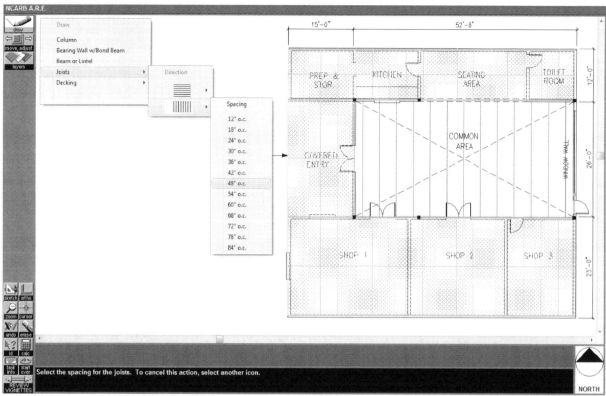

Figure 4.24   Use **Draw > Joists > Select joists Direction > Spacing at 48" o.c.** to draw joists between the bearing beams.

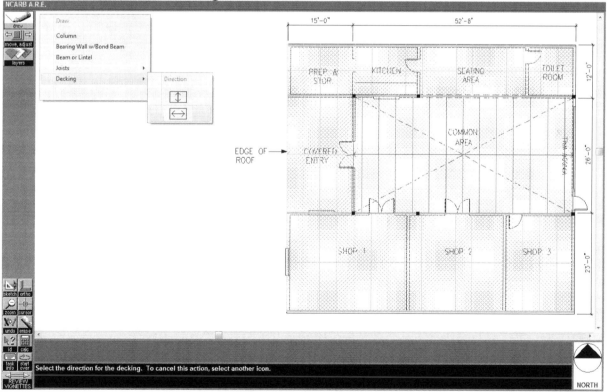

Figure 4.25   Use **Draw > Decking > Select deck direction** to draw decks between the joists.

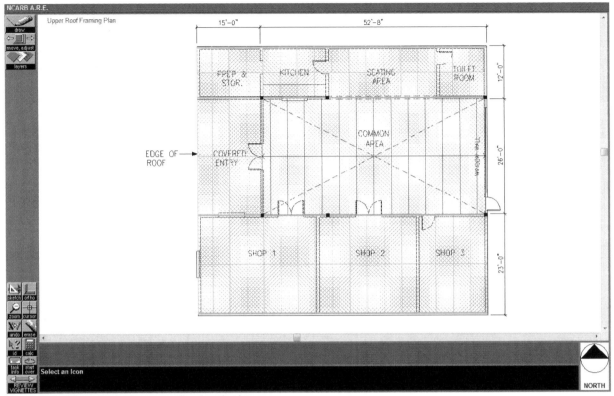

Figure 4.26   This is your final solution.

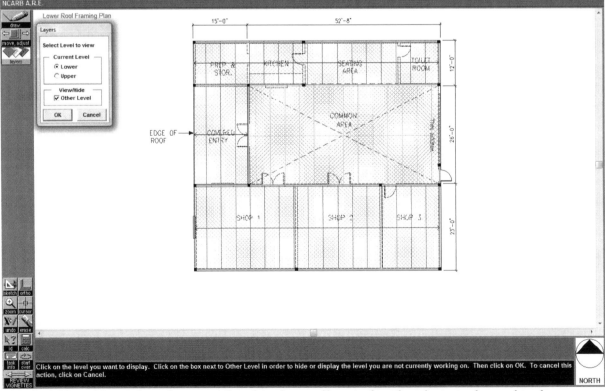

Figure 4.27   Use only columns and no bearing walls to solve the vignette: lower level.

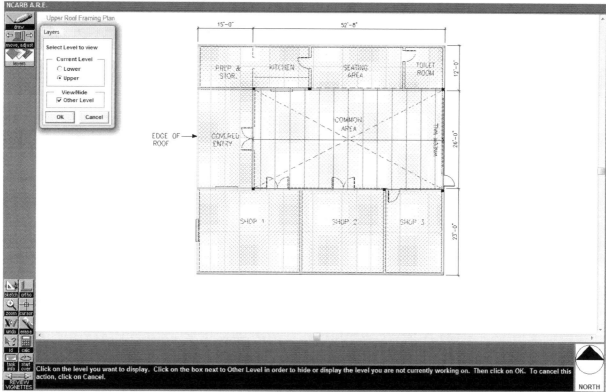

Figure 4.28   Use only columns and no bearing walls to solve the vignette: upper level.

Appendixes

A. List of Figures

B. Official reference materials suggested by NCARB

1. General NCARB reference materials for ARE:

Per NCARB, all candidates should become familiar with the latest version of the following codes:

International Code Council, Inc. (ICC, 2006)
International Building Code
International Mechanical Code
International Plumbing Code

National Fire Protection Association (NFPA)
Life Safety Code (NFPA 101)
National Electrical Code (NFPA 70)

National Research Council of Canada
National Building Code of Canada
National Plumbing Code of Canada
National Fire Code of Canada

American Institute of Architects
AIA Documents - 2007

Candidates should be familiar with the Standard on Accessible and Usable Buildings and Facilities (ICC/ANSI A117.1-98)

2. Official NCARB reference materials for the Structural Systems (SS) division:

ACI Code 318-05 (Building Code Requirements for Reinforced Concrete)
American Concrete Institute, 2005

OR
CAN/CSA-A23.1-94 (Concrete Materials and Methods of Concrete Construction) and CAN/CSA-A23.3-94 (Design of Concrete Structures for Buildings)
Canadian Standards Association

Architectural Graphic Standards
Charles G. Ramsey and Harold R. Sleeper
The American Institute of Architects
John Wiley & Sons, latest edition

Building Structures
James Ambrose
John Wiley & Sons, 1993

Design Value for Wood Construction
American Wood Council, 2005

Elementary Structures for Architects and Builders, Fourth Edition
Ronald E. Shaeffer
Prentice Hall, 2006

Introduction to Wood Design
Canadian Wood Council, 2005

Manual of Steel Construction: Allowable Stress Design; 9th Edition.
American Institute of Steel Construction, Inc. Chicago, Illinois, 1989

National Building Code of Canada, 2005
Parts 1, 3, 4, 9; Appendix A
Supplement
Chapters 1, 2, 4; Commentaries A, D, F, H, I

NEHRP (National Earthquake Hazards Reduction Program) Recommended Provisions for Seismic Regulations for New Buildings and Other Structures Parts 1 and 2
FEMA 2003

Simplified Building Design for Wind and Earthquake Forces
James Ambrose and Dimitry Vergun
John Wiley & Sons, 1997

Simplified Design of Concrete Structures,
Eighth Edition
James Ambrose, Patrick Tripeny
John Wiley & Sons, 2007

Simplified Design of Masonry Structures
James Ambrose
John Wiley & Sons, 1997

Simplified Design of Steel Structures, Eighth Edition
James Ambrose, Patrick Tripeny
John Wiley & Sons, 2007

Simplified Design of Wood Structures, Fifth Edition
James Ambrose
John Wiley & Sons, 2009

Simplified Engineering for Architects and Builders, Tenth Edition
Harry Parker and James Ambrose
John Wiley & Sons, 2005

Simplified Mechanics and Strength of Materials, Fifth Edition
Harry Parker and James Ambrose
John Wiley & Sons, 2002

Standard Specifications Load Tables & Weight Tables for Steel Joists and Joist Girders
Steel Joist Institute, latest edition

Steel Construction Manual, Latest edition
American Institute of Steel Construction, 2006

OR
Handbook of Steel Construction, Latest edition; and *CAN/CSA-S16-01 and CISC Commentary*
Canadian Institute of Steel Construction

Steel Deck Institute Tables
Steel Deck Institute

OR
LSD Steel Deck Tables
Caradon Metal Building Products

Structural Concepts and Systems for Architects and Engineers, Second Edition
T.Y. Lin and Sidney D. Stotesbury
Van Nostrand Reinhold, 1988

Structural Design: A Practical Guide for Architects
James Underwood and Michele Chiuini
John Wiley & Sons, latest edition

Structure in Architecture: The Building of Buildings
Mario Salvadori with Robert Heller
Prentice-Hall, 1986

Understanding Structures
Fuller Moore
McGraw-Hill, 1999

Wood Design Manual and *CAN/CSA-086.1-94 and Commentary*
Canadian Wood Council

3. Official NCARB list of formulas and references for the Structural Systems (SS) division:
You can download the NCARB list of formulas and references for the Structural Systems (SS) division at the following link:
http://www.ncarb.org/en/ARE/Preparing-for-the-ARE.aspx

These formulas and references will be available during the real exam. You should read through them a few times before the exam to become familiar with them. This will save you a lot of time during the real exam, and will help you solve structural calculations and other problems.

C. Other reference materials

Chen, Gang. *Building Construction: Project Management, Construction Administration, Drawings, Specs, Detailing Tips, Schedules, Checklists, and Secrets Others Don't Tell You (Architectural Practice Simplified, 2nd edition).* ArchiteG, Inc., A good introduction to the architectural practice and construction documents and service, including discussions of MasterSpec format and specification sections.

Chen, Gang. *LEED GA Exam Guide: A Must-Have for the LEED Green Associate Exam: Comprehensive Study Materials, Sample Questions, Mock Exam, Green Building LEED Certification, and Sustainability.* ArchiteG, Inc., latest edition. A good introduction to green buildings and the LEED building rating system.

Ching, Francis. *Architecture: Form, Space, & Order.* Wiley, latest edition. It is one of the best architectural books that you can have. I still flip through it every now and then. It is a great book for inspiration.

Ching, Francis. Steven R. Winkel, FAIA, PE. *Building Codes Illustrated: A Guide to Understanding the International Building Code.* Wiley, latest edition. A valuable interpretive guide with many useful line drawings. A great timesaver.

Frampton, Kenneth. *Modern Architecture: A Critical History.* Thames and Hudson, London, latest edition. A valuable resource for architectural history.

Jarzombek, Mark M. (Author), Vikramaditya Prakash (Author), Francis D. K. Ching (Editor). *A Global History of Architecture.* Wiley, latest edition. A valuable and comprehensive resource for architectural history with 1000 b & w photos, 50 color photos, and 1500 b & w illustrations. It doesn't limit the topic on a Western perspective, but rather through a global vision.

Trachtenberg, Marvin and Isabelle Hyman. *Architecture: From Pre-history to Post-Modernism.* Prentice Hall, Englewood Cliffs, NJ   latest edition.   A valuable and comprehensive resource for architectural history.

D. Definition of Architects and Some Important Information about Architects and the Profession of Architecture

Architects, Except Landscape and Naval

- Nature of the Work
- Training, Other Qualifications, and Advancement
- Employment
- Job Outlook
- Projections Data
- Earnings
- OES Data
- Related Occupations
- Sources of Additional Information

Significant Points

- About 1 in 5 architects are self-employed—more than 2 times the proportion for all occupations.
- Licensing requirements include a professional degree in architecture, at least 3 years of practical work training, and passing all divisions of the Architect Registration Examination.
- Architecture graduates may face competition, especially for jobs in the most prestigious firms.

Nature of the Work

People need places in which to live, work, play, learn, worship, meet, govern, shop, and eat. These places may be private or public; indoors or out; rooms, buildings, or complexes, and architects design them. Architects are licensed professionals trained in the art and science of building design who develop the concepts for structures and turn those concepts into images and plans.

Architects create the overall aesthetic and look of buildings and other structures, but the design of a building involves far more than its appearance. Buildings also must be functional, safe, and economical and must suit the needs of the people who use them. Architects consider all these factors when they design buildings and other structures.

Architects may be involved in all phases of a construction project, from the initial discussion with the client through the entire construction process. Their duties require specific skills—designing, engineering, managing, supervising, and communicating with clients and builders. Architects spend a great deal of time explaining their ideas to clients, construction contractors, and others. Successful architects must be able to communicate their unique vision persuasively.

The architect and client discuss the objectives, requirements, and budget of a project. In some cases, architects provide various pre-design services: conducting feasibility and environmental impact studies, selecting a site, preparing cost analysis and land-use studies, or specifying the requirements the design must meet. For example, they may determine space requirements by researching the numbers and types of potential users of a building. The architect then prepares drawings and a report presenting ideas for the client to review. After discussing and agreeing on the initial proposal, architects develop final construction plans that show the building's appearance and details for its construction. Accompanying these plans are drawings of the structural system; air-conditioning, heating, and ventilating systems; electrical systems; communications systems; plumbing; and, possibly, site and landscape plans. The plans also specify the building materials and, in some cases, the interior furnishings. In developing designs, architects follow building codes, zoning laws, fire regulations, and other ordinances, such as those requiring easy access by people who are disabled. Computer-aided design and drafting (CADD) and Building Information Modeling (BIM) technology has replaced traditional paper and pencil as the most common method for creating design and construction drawings. Continual revision of plans on the basis of client needs and budget constraints is often necessary.

Architects may also assist clients in obtaining construction bids, selecting contractors, and negotiating construction contracts. As construction proceeds, they may visit building sites to make sure that contractors follow the design, adhere to the schedule, use the specified materials, and meet work quality standards. The job is not complete until all construction is finished, required tests are conducted, and construction costs are paid. Sometimes, architects also provide post-construction services, such as facilities management. They advise on energy efficiency measures, evaluate how well the building design adapts to the needs of occupants, and make necessary improvements.

Often working with engineers, urban planners, interior designers, landscape architects, and other professionals, architects in fact spend a great deal of their time coordinating information from, and the work of, other professionals engaged in the same project.

They design a wide variety of buildings, such as office and apartment buildings, schools, churches, factories, hospitals, houses, and airport terminals. They also design complexes such as urban centers, college campuses, industrial parks, and entire communities.

Architects sometimes specialize in one phase of work. Some specialize in the design of one type of building—for example, hospitals, schools, or housing. Others focus on planning and pre-design services or construction management and do minimal design work.

Work environment. Usually working in a comfortable environment, architects spend most of their time in offices consulting with clients, developing reports and drawings, and working with other architects and engineers. However, they often visit construction sites to review the progress of projects. Although most architects work approximately 40 hours per week, they often have to work nights and weekends to meet deadlines.

Training, Other Qualifications, and Advancement

There are three main steps in becoming an architect. First is the attainment of a professional degree in architecture. Second is work experience through an internship, and third is licensure through the passing of the Architect Registration Exam.

Education and training. In most States, the professional degree in architecture must be from one of the 114 schools of architecture that have degree programs accredited by the National Architectural Accrediting Board. However, State architectural registration boards set their own standards, so graduation from a non-accredited program may meet the educational requirement for licensing in a few States.

Three types of professional degrees in architecture are available: a 5-year bachelor's degree, which is most common and is intended for students with no previous architectural training; a 2-year master's degree for students with an undergraduate degree in architecture or a related area; and a 3- or 4-year master's degree for students with a degree in another discipline.

The choice of degree depends on preference and educational background. Prospective architecture students should consider the options before committing to a program. For example, although the 5-year bachelor of architecture offers the fastest route to the professional degree, courses are specialized, and if the student does not complete the program, transferring to a program in another discipline may be difficult. A typical program includes courses in architectural history and theory, building design with an emphasis on CADD, structures, technology, construction methods, professional practice, math, physical sciences, and liberal arts. Central to most architectural programs is the design studio, where students apply the skills and concepts learned in the classroom, creating drawings and three-dimensional models of their designs.

Many schools of architecture also offer post-professional degrees for those who already have a bachelor's or master's degree in architecture or other areas. Although graduate education beyond the professional degree is not required for practicing architects, it may be required for research, teaching, and certain specialties.

All State architectural registration boards require architecture graduates to complete a training period—usually at least 3 years—before they may sit for the licensing exam. Every State, with the exception of Arizona, has adopted the training standards established by the Intern Development Program, a branch of the American Institute of Architects and the National Council of Architectural Registration Boards (NCARB). These standards stipulate broad training under the supervision of a licensed architect. Most new graduates complete their training period by working as interns at architectural firms. Some States allow a portion of the training to occur in the offices of related professionals, such as engineers or general contractors. Architecture students who complete internships while still in school can count some of that time toward the 3-year training period.

Interns in architectural firms may assist in the design of one part of a project, help prepare architectural documents or drawings, build models, or prepare construction drawings on CADD. Interns also may research building codes and materials or write specifications for building materials, installation criteria, the quality of finishes, and other, related details.

Licensure. All States and the District of Columbia require individuals to be licensed (registered) before they may call themselves architects and contract to provide architectural services. During the time between graduation and becoming licensed, architecture school graduates generally work in the field under the supervision of a licensed architect who takes legal responsibility for all work. Licensing requirements include a professional degree in architecture, a period of practical training or internship, and a passing score on all divisions of the Architect Registration Examination. The examination is broken into nine divisions consisting of either multiple choice or graphical questions. The eligibility period for completion of all divisions of the exam varies by State.

Most States also require some form of continuing education to maintain a license, and many others are expected to adopt mandatory continuing education. Requirements vary by State but usually involve the completion of a certain number of credits annually or biennially through workshops, formal university classes, conferences, self-study courses, or other sources.

Other qualifications. Architects must be able to communicate their ideas visually to their clients. Artistic and drawing ability is helpful, but not essential, to such communication. More important are a visual orientation and the ability to understand spatial relationships. Other important qualities for anyone interested in becoming an architect are creativity and the ability to work independently and as part of a team. Computer skills are also required for writing specifications, for 2- and 3- dimensional drafting using CADD programs, and for financial management.

Certification and advancement. A growing number of architects voluntarily seek certification by the National Council of Architectural Registration Boards. Certification is awarded after independent verification of the candidate's educational transcripts, employment record, and professional references. Certification can make it easier to become licensed across States. In fact, it is the primary requirement for reciprocity of licensing among State Boards that are NCARB members. In 2007, approximately one-third of all licensed architects had this certification.

After becoming licensed and gaining experience, architects take on increasingly responsible duties, eventually managing entire projects. In large firms, architects may advance to supervisory or managerial positions. Some architects become partners in established firms, while others set up their own practices. Some graduates with degrees in architecture also enter related fields, such as graphic, interior, or industrial design; urban planning; real estate development; civil engineering; and construction management.

Employment

Architects held about 132,000 jobs in 2006. Approximately 7 out of 10 jobs were in the architectural, engineering, and related services industry—mostly in architectural firms with fewer than five workers. A small number worked for residential and nonresidential building construction firms and for government agencies responsible for housing, community planning, or construction of government buildings, such as the U.S. Departments of Defense and Interior, and the General Services Administration. About 1 in 5 architects are self-employed.

Job Outlook

Employment of architects is expected to grow faster than the average for all occupations through 2016. Keen competition is expected for positions at the most prestigious firms, and opportunities will be best for those architects who are able to distinguish themselves with their creativity.

Employment change. Employment of architects is expected to grow by 18 percent between 2006 and 2016, which is faster than the average for all occupations. Employment of architects is strongly tied to the activity of the construction industry. Strong growth is expected to come from nonresidential construction as demand for commercial space increases. Residential construction, buoyed by low interest rates, is also expected to grow as more people become homeowners. If interest rates rise significantly, home building may fall off, but residential construction makes up only a small part of architects' work.

Current demographic trends also support an increase in demand for architects. As the population of Sunbelt States continues to grow, the people living there will need new places to live and work. As the population continues to live longer and baby-boomers begin to retire, there will be a need for more healthcare facilities, nursing homes, and retirement communities. In education, buildings at all levels are getting older and class sizes are getting larger. This will require many school districts and universities to build new facilities and renovate existing ones.

In recent years, some architecture firms have outsourced the drafting of construction documents and basic design for large-scale commercial and residential projects to architecture firms overseas. This trend is expected to continue and may have a negative impact on employment growth for lower level architects and interns who would normally gain experience by producing these drawings.

Job prospects. Besides employment growth, additional job openings will arise from the need to replace the many architects who are nearing retirement, and others who transfer to other occupations or stop working for other reasons. Internship opportunities for new architectural students are expected to be good over the next decade, but more students are graduating with architectural degrees and some competition for entry-level jobs can be anticipated. Competition will be especially keen for jobs at the most prestigious architectural firms as prospective architects try to build their reputation. Prospective architects who have had internships while in school will have an advantage in obtaining intern positions after graduation. Opportunities will be best for those architects that are able to distinguish themselves from others with their creativity.

Prospects will also be favorable for architects with knowledge of "green" design. Green design, also known as sustainable design, emphasizes energy efficiency, renewable resources such as energy and water, waste reduction, and environmentally friendly design, specifications, and materials. Rising energy costs and increased concern about the environment has led to many new buildings being built green.

Some types of construction are sensitive to cyclical changes in the economy. Architects seeking design projects for office and retail construction will face especially strong competition for jobs or clients during recessions, and layoffs may ensue in less successful firms. Those involved in the design of institutional buildings, such as schools, hospitals, nursing homes, and correctional facilities, will be less affected by fluctuations in the economy. Residential construction makes up a small portion of work for architects, so major changes in the housing market would not be as significant as fluctuations in the nonresidential market.

Despite good overall job opportunities, some architects may not fare as well as others. The profession is geographically sensitive, and some parts of the Nation may have fewer new building projects. Also, many firms specialize in specific buildings, such as hospitals or office towers, and demand for these buildings may vary by region. Architects may find it increasingly necessary to gain reciprocity in order to compete for the best jobs and projects in other States.

Projections Data

Projections data from the National Employment Matrix

Occupational title	SOC Code	Employment, 2006	Projected employment, 2016	Change, 2006-16		Detailed statistics
				Number	Percent	
Architects, except landscape and naval	17-1011	132,000	155,000	23,000	18	PDF zipped XLS

NOTE: Data in this table are rounded. See the discussion of the employment projections table in the *Handbook* introductory chapter on *Occupational Information Included in the Handbook*.

Earnings

Median annual earnings of wage-and-salary architects were $64,150 in May 2006. The middle 50 percent earned between $49,780 and $83,450. The lowest 10 percent earned less than $39,420, and the highest 10 percent earned more than $104,970. Those just starting their internships can expect to earn considerably less.

Earnings of partners in established architectural firms may fluctuate because of changing business conditions. Some architects may have difficulty establishing their own practices and may go through a period when their expenses are greater than their income, requiring substantial financial resources.

Many firms pay tuition and fees toward continuing education requirements for their employees.

For the latest wage information:

The above wage data are from the Occupational Employment Statistics (OES) survey program, unless otherwise noted. For the latest National, State, and local earnings data, visit the following pages:

Architects, except landscape and naval

Related Occupations

Architects design buildings and related structures. Construction managers, like architects, also plan and coordinate activities concerned with the construction and maintenance of buildings and facilities. Others who engage in similar work are landscape architects, civil engineers, urban and regional planners, and designers, including interior designers, commercial and industrial designers, and graphic designers.

Sources of Additional Information

Disclaimer:

Links to non-BLS Internet sites are provided for your convenience and do not constitute an endorsement.

Information about education and careers in architecture can be obtained from:

- The American Institute of Architects, 1735 New York Ave. NW., Washington, DC 20006. Internet: http://www.aia.org
- Intern Development Program, National Council of Architectural Registration Boards, Suite 1100K, 1801 K St. NW., Washington, D.C. 20006. Internet: http://www.ncarb.org OOH ONET Codes 17-1011.00"

Quoted from: Bureau of Labor Statistics, U.S. Department of Labor, Occupational Outlook Handbook, 2008-09 Edition, Architects, Except Landscape and Naval, on the Internet at **http://www.bls.gov/oco/ocos038.htm** (visited November 30, 2008).

Last Modified Date: December 18, 2007

Note: Please check the website above for the latest information.

E. AIA Compensation Survey

Every 3 years, AIA publishes a Compensation Survey for various positions at architectural firms across the country. It is a good idea to find out the salary before you make the final decision to become an architect. If you are already an architect, it is also a good idea to determine if you are underpaid or overpaid.

See following link for some sample pages for the 2008 AIA Compensation Survey:

http://www.aia.org/aiaucmp/groups/ek_public/documents/pdf/aiap072881.pdf

F.   So … You would Like to Study Architecture

To study architecture, you need to learn how to draft, how to understand and organize spaces and the interactions between interior and exterior spaces, how to do design, and how to communicate effectively. You also need to understand the history of architecture.

As an architect, a leader for a team of various design professionals, you not only need to know architecture, but also need to understand enough of your consultants' work to be able to coordinate them. Your consultants include soils and civil engineers, landscape architects, structural, electrical, mechanical, and plumbing engineers, interior designers, sign consultants, etc.

There are two major career paths for you in architecture: practice as an architect or teach in colleges or universities. The earlier you determine which path you are going to take, the more likely you will be successful at an early age. Some famous and well-respected architects, like my USC alumnus Frank Gehry, have combined the two paths successfully. They teach at the universities and have their own architectural practice. Even as a college or university professor, people respect you more if you have actual working experience and have some built projects. If you only teach in colleges or universities but have no actual working experience and have no built projects, people will consider you as a "paper" architect, and they are not likely to take you seriously, because they will think you probably do not know how to put a real building together.

In the U.S., if you want to practice architecture, you need to obtain an architect's license. It requires a combination of passing scores on the Architectural Registration Exam (ARE) and 8 years of education and/or qualified working experience, including at least 1 year of working experience in the U.S. Your working experience needs to be under the supervision of a licensed architect to be counted as qualified working experience for your architect's license.

If you work for a landscape architect or civil engineer or structural engineer, some states' architectural licensing boards will count your experience at a discounted rate for the qualification of your architect's license. For example, 2 years of experience working for a civil engineer may be counted as 1 year of qualified experience for your architect's license. You need to contact your state's architectural licensing board for specific licensing requirements for your state.

If you want to teach in colleges or universities, you probably want to obtain a master's degree or a Ph.D. It is not very common for people in the architectural field to have a Ph.D. One reason is that there are few Ph.D. programs for architecture. Another reason is that architecture is considered a profession and requires a license. Many people think an architect's license is more important than a Ph.D. degree. In many states, you need to have an architect's license to even use the title "architect," or the terms "architectural" or "architecture" to advertise your service. You cannot call yourself an architect if you do not have an architect's license, even if you have a Ph.D. in architecture. Violation of these rules brings punishment.

To become a tenured professor, you need to have a certain number of publications and pass the evaluation for the tenure position. Publications are very important for tenure track positions. Some people say for the tenured track positions in universities and colleges, it is "publish or perish."

The American Institute of Architects (AIA) is the national organization for the architectural profession. Membership is voluntary. There are different levels of AIA membership. Only licensed architects can be (full) AIA members. If you are an architectural student or an intern but not a licensed architect yet, you can join as an associate AIA member. Contact AIA for detailed information.

The National Council of Architectural Registration Boards (NCARB) is a nonprofit federation of architectural licensing boards. It has some very useful programs, such as IDP, to assist you in obtaining your architect's license. Contact NCARB for detailed information.

Back Page Promotion

You may be interested in some other books written by Gang Chen:

A. **ARE Mock Exam series. See the following link:**
   http://www.GreenExamEducation.com

B. **LEED Exam Guides series. See the following link:**
   http://www.GreenExamEducation.com

C. ***Building Construction:*** *Project Management, Construction Administration, Drawings, Specs, Detailing Tips, Schedules, Checklists, and Secrets Others Don't Tell You (Architectural Practice Simplified, 2nd edition)*
   http://www.ArchiteG.com

D. ***Planting Design Illustrated***
   http://outskirtspress.com/agent.php?key=11011&page=GangChen

ARE Mock Exam Series

Published ARE books (One Mock Exam book for each ARE division, plus California Supplemental Mock Exam):
Programming, Planning & Practice (PPP) ARE Mock Exam (Architect Registration Exam): ARE Overview, Exam Prep Tips, Multiple-Choice Questions and Graphic Vignettes, Solutions and Explanations. **ISBN-13:** 9781612650067

Site Planning & Design ARE Mock Exam (SPD of Architect Registration Exam): ARE Overview, Exam Prep Tips, Multiple-Choice Questions and Graphic Vignettes, Solutions and Explanations. **ISBN-13:** 9781612650111

Building Design and Construction Systems (BDCS) ARE Mock Exam (Architect Registration Exam): ARE Overview, Exam Prep Tips, Multiple-Choice Questions and Graphic Vignettes, Solutions and Explanations. **ISBN-13:** 9781612650029

Schematic Design (SD) ARE Mock Exam (Architect Registration Exam): ARE Overview, Exam Prep Tips, Graphic Vignettes, Solutions and Explanations
ISBN: 9781612650050

Structural Systems ARE Mock Exam (SS of Architect Registration Exam): ARE Overview, Exam Prep Tips, Multiple-Choice Questions and Graphic Vignettes, Solutions and Explanations. **ISBN**: 9781612650012

Building Systems (BS) ARE Mock Exam (Architect Registration Exam): ARE Overview, Exam Prep Tips, Multiple-Choice Questions and Graphic Vignettes, Solutions and Explanations. **ISBN-13**: 9781612650036

Construction Documents and Service (CDS) Are Mock Exam (Architect Registration Exam): ARE Overview, Exam Prep Tips, Multiple-Choice Questions and Graphic Vignettes, Solutions and Explanations. **ISBN-13:** 9781612650005

Mock California Supplemental Exam (CSE of Architect Registration Exam): CSE Overview, Exam Prep Tips, General Section and Project Scenario Section, Questions, Solutions and Explanations. **ISBN**: 9781612650159

Upcoming ARE books:
Other books in the ARE Mock Exam Series are being produced. Our goal is to produce one mock exam book PLUS one guidebook for each of the ARE exam divisions.

See the following link for the latest information:
http://www.GreenExamEducation.com

LEED Exam Guides series*: Comprehensive Study Materials, Sample Questions, Mock Exam, Building LEED Certification and Going Green**

LEED (Leadership in Energy and Environmental Design) is the most important trend of development, and it is revolutionizing the construction industry. It has gained tremendous momentum and has a profound impact on our environment.

From LEED Exam Guides series, you will learn how to

1. Pass the LEED Green Associate Exam and various LEED AP + exams (each book will help you with a specific LEED exam).

2. Register and certify a building for LEED certification.

3. Understand the intent for each LEED prerequisite and credit.

4. Calculate points for a LEED credit.

5. Identify the responsible party for each prerequisite and credit.

6. Earn extra credit (exemplary performance) for LEED.

7. Implement the local codes and building standards for prerequisites and credit.

8. Receive points for categories not yet clearly defined by USGBC.

There is currently NO official book on the LEED Green Associate Exam, and most of the existing books on LEED and LEED AP are too expensive and too complicated to be practical and helpful. The pocket guides in LEED Exam Guides series fill in the blanks, demystify LEED, and uncover the tips, codes, and jargon for LEED as well as the true meaning of "going green." They will set up a solid foundation and fundamental framework of LEED for you. Each book in the LEED Exam Guides series covers every aspect of one or more specific LEED rating system(s) in plain and concise language and makes this information understandable to all people.

These pocket guides are small and easy to carry around. You can read them whenever you have a few extra minutes. They are indispensable books for all people—administrators; developers; contractors; architects; landscape architects; civil, mechanical, electrical, and plumbing engineers; interns; drafters; designers; and other design professionals.

Why is the LEED Exam Guides series needed?

A number of books are available that you can use to prepare for the LEED exams:

1. *USGBC Reference Guides.* You need to select the correct version of the *Reference Guide* for your exam.

   The *USGBC Reference Guides* are comprehensive, but they give too much information. For example, *The LEED 2009 Reference Guide for Green Building Design and Construction (BD&C)* has about 700 oversized pages. Many of the calculations in the books are too detailed for the exam. They are also expensive (approximately $200 each, so most people may not buy them for their personal use, but instead, will seek to share an office copy).

   It is good to read a reference guide from cover to cover if you have the time. The problem is not too many people have time to read the whole reference guide. Even if you do read the whole guide, you may not remember the important issues to pass the LEED exam. You need to reread the material several times before you can remember much of it.

   Reading the reference guide from cover to cover without a guidebook is a difficult and inefficient way of preparing for the LEED AP Exam, because you do NOT know what USGBC and GBCI are looking for in the exam.

2. The USGBC workshops and related handouts are concise, but they do not cover extra credits (exemplary performance). The workshops are expensive, costing approximately $450 each.

3. Various books published by a third party are available on Amazon, bn.com and books.google.com. However, most of them are not very helpful.

   There are many books on LEED, but not all are useful.

   LEED Exam Guides series will fill in the blanks and become a valuable, reliable source:

   a. They will give you more information for your money. Each of the books in the LEED Exam Guides series has more information than the related USGBC workshops.

   b. They are exam-oriented and more effective than the USGBC reference guides.

   c. They are better than most, if not all, of the other third-party books. They give you comprehensive study materials, sample questions and answers, mock exams and answers, and critical information on building LEED certification and going green. Other third-party books only give you a fraction of the information.

   d. They are comprehensive yet concise. They are small and easy to carry around. You can read them whenever you have a few extra minutes.

   e. They are great timesavers. I have highlighted the important information that you need to understand and MEMORIZE. I also make some acronyms and short sentences to help you easily remember the credit names.

It should take you about 1 or 2 weeks of full-time study to pass each of the LEED exams. I have met people who have spent 40 hours to study and passed the exams.

You can find sample texts and other information on the LEED Exam Guides series in customer discussion sections under each of my book's listing on Amazon, bn.com and books.google.com.

What others are saying about *LEED GA Exam Guide* (Book 2, LEED Exam Guide series):

"Finally! A comprehensive study tool for LEED GA Prep!

"I took the 1-day Green LEED GA course and walked away with a power point binder printed in very small print—which was missing MUCH of the required information (although I didn't know it at the time). I studied my little heart out and took the test, only to fail it by 1 point. Turns out I did NOT study all the material I needed to in order to pass the test. I found this book, read it, marked it up, retook the test, and passed it with a 95%. Look, we all know the LEED GA exam is new and the resources for study are VERY limited. This one is the VERY best out there right now. I highly recommend it."
—ConsultantVA

"Complete overview for the LEED GA exam

"I studied this book for about 3 days and passed the exam … if you are truly interested in learning about the LEED system and green building design, this is a great place to start."
—K.A. Evans

"A Wonderful Guide for the LEED GA Exam

"After deciding to take the LEED Green Associate exam, I started to look for the best possible study materials and resources. From what I thought would be a relatively easy task, it turned into a tedious endeavor. I realized that there are vast amounts of third-party guides and handbooks. Since the official sites offer little to no help, it became clear to me that my best chance to succeed and pass this exam would be to find the most comprehensive study guide that would not only teach me the topics, but would also give me a great background and understanding of what LEED actually is. Once I stumbled upon Mr. Chen's book, all my needs were answered. This is a great study guide that will give the reader the most complete view of the LEED exam and all that it entails.

"The book is written in an easy-to-understand language and brings up great examples, tying the material to the real world. The information is presented in a coherent and logical way, which optimizes the learning process and does not go into details that will not be needed for the LEED Green Associate Exam, as many other guides do. This book stays dead on topic and keeps the reader interested in the material.

"I highly recommend this book to anyone that is considering the LEED Green Associate Exam. I learned a great deal from this guide, and I am feeling very confident about my chances for passing my upcoming exam."
—Pavel Geystrin

"Easy to read, easy to understand

"I have read through the book once and found it to be the perfect study guide for me. The author does a great job of helping you get into the right frame of mind for the content of the exam. I had started by studying the Green Building Design and Construction reference guide for LEED projects produced by the USGBC. That was the wrong approach, simply too much information with very little retention. At 636 pages in textbook format, it would have been a daunting task to get through it. Gang Chen breaks down the points, helping to minimize the amount of information but maximizing the content I was able to absorb. I plan on going through the book a few more times, and I now believe I have the right information to pass the LEED Green Associate Exam."
—**Brian Hochstein**

"All in one—LEED GA prep material

"Since the LEED Green Associate exam is a newer addition by USGBC, there is not much information regarding study material for this exam. When I started looking around for material, I got really confused about what material I should buy. This LEED GA guide by Gang Chen is an answer to all my worries! It is a very precise book with lots of information, like how to approach the exam, what to study and what to skip, links to online material, and tips and tricks for passing the exam. It is like the 'one stop shop' for the LEED Green Associate Exam. I think this book can also be a good reference guide for green building professionals. A must-have!"
—**SwatiD**

"An ESSENTIAL LEED GA Exam Reference Guide

"This book is an invaluable tool in preparation for the LEED Green Associate (GA) Exam. As a practicing professional in the consulting realm, I found this book to be all-inclusive of the preparatory material needed for sitting the exam. The information provides clarity to the fundamental and advanced concepts of what LEED aims to achieve. A tremendous benefit is the connectivity of the concepts with real-world applications.

"The author, Gang Chen, provides a vast amount of knowledge in a very clear, concise, and logical media. For those that have not picked up a textbook in a while, it is very manageable to extract the needed information from this book. If you are taking the exam, do yourself a favor and purchase a copy of this great guide. Applicable fields: Civil Engineering, Architectural Design, MEP, and General Land Development."
—**Edwin L. Tamang**

Note: Other books in the **LEED Exam Guides series** are in the process of being produced. At least **one book will eventually be produced for each of the LEED exams.** The series include:

LEED v4 Green Associate Exam Guide (LEED GA): *Comprehensive Study Materials, Sample Questions, Mock Exam, Green Building LEED Certification, and Sustainability*, LEED Exam Guide series, ArchiteG.com. Latest Edition.

LEED GA MOCK EXAMS (LEED v4): *Questions, Answers, and Explanations: A Must-Have for the LEED Green Associate Exam, Green Building LEED Certification, and Sustainability,* LEED Exam Guide series, ArchiteG.com. Latest Edition

LEED v4 BD&C EXAM GUIDE: *A Must-Have for the LEED AP BD+C Exam: Comprehensive Study Materials, Sample Questions, Mock Exam, Green Building Design and Construction, LEED Certification, and Sustainability,* LEED Exam Guide series, ArchiteG.com. Latest Edition.

LEED v4 BD&C MOCK EXAMS: *Questions, Answers, and Explanations: A Must-Have for the LEED AP BD+C Exam, Green Building LEED Certification, and Sustainability,* LEED Exam Guide series, ArchiteG.com. Latest Edition.

LEED ID&C Exam Guide: *A Must-Have for the LEED AP ID+C Exam: Study Materials, Sample Questions, Green Interior Design and Construction, Green Building LEED Certification, and Sustainability,* LEED Exam Guide series, ArchiteG.com. Latest Edition.

LEED ID&C Mock Exam: *Questions, Answers, and Explanations: A Must-Have for the LEED AP ID+C Exam, Green Interior Design and Construction, Green Building LEED Certification, and Sustainability,* LEED Exam Guide series, ArchiteG.com. Latest Edition.

LEED O&M MOCK EXAMS: *Questions, Answers, and Explanations: A Must-Have for the LEED O&M Exam, Green Building LEED Certification, and Sustainability,* LEED Exam Guide series, ArchiteG.com. Latest Edition.

LEED O&M EXAM GUIDE: *A Must-Have for the LEED AP O+M Exam: Comprehensive Study Materials, Sample Questions, Mock Exam, Green Building Operations and Maintenance, LEED Certification, and Sustainability,* LEED Exam Guide series, ArchiteG.com. Latest Edition.

LEED HOMES EXAM GUIDE: *A Must-Have for the LEED AP Homes Exam: Comprehensive Study Materials, Sample Questions, Mock Exam, Green Building LEED Certification, and Sustainability,* LEED Exam Guide series, ArchiteG.com. Latest Edition.

LEED ND EXAM GUIDE: *A Must-Have for the LEED AP Neighborhood Development Exam: Comprehensive Study Materials, Sample Questions, Mock Exam, Green Building LEED Certification, and Sustainability,* LEED Exam Guide series, ArchiteG.com. Latest Edition.

How to order these books:
You can order the books listed above at:
http://www.GreenExamEducation.com

OR
http://www.ArchiteG.com

Building Construction

Project Management, Construction Administration, Drawings, Specs, Detailing Tips, Schedules, Checklists, and Secrets Others Don't Tell You (Architectural Practice Simplified, 2nd edition)

Learn the Tips, Become One of Those Who Know Building Construction and Architectural Practice, and Thrive!

For architectural practice and building design and construction industry, there are two kinds of people: those who know, and those who don't. The tips of building design and construction and project management have been undercover—until now.

Most of the existing books on building construction and architectural practice are too expensive, too complicated, and too long to be practical and helpful. This book simplifies the process to make it easier to understand and uncovers the tips of building design and construction and project management. It sets up a solid foundation and fundamental framework for this field. It covers every aspect of building construction and architectural practice in plain and concise language and introduces it to all people. Through practical case studies, it demonstrates the efficient and proper ways to handle various issues and problems in architectural practice and building design and construction industry.

It is for ordinary people and aspiring young architects as well as seasoned professionals in the construction industry. For ordinary people, it uncovers the tips of building construction; for aspiring architects, it works as a construction industry survival guide and a guidebook to shorten the process in mastering architectural practice and climbing up the professional ladder; for seasoned architects, it has many checklists to refresh their memory. It is an indispensable reference book for ordinary people, architectural students, interns, drafters, designers, seasoned architects, engineers, construction administrators, superintendents, construction managers, contractors, and developers.

You will learn:
1. How to develop your business and work with your client.
2. The entire process of building design and construction, including programming, entitlement, schematic design, design development, construction documents, bidding, and construction administration.
3. How to coordinate with governing agencies, including a county's health department and a city's planning, building, fire, public works departments, etc.
4. How to coordinate with your consultants, including soils, civil, structural, electrical, mechanical, plumbing engineers, landscape architects, etc.
5. How to create and use your own checklists to do quality control of your construction documents.
6. How to use various logs (i.e., RFI log, submittal log, field visit log, etc.) and lists (contact list, document control list, distribution list, etc.) to organize and simplify your work.
7. How to respond to RFI, issue CCDs, review change orders, submittals, etc.
8. How to make your architectural practice a profitable and successful business.

Planting Design Illustrated
A Must-Have for Landscape Architecture: A Holistic Garden Design Guide with Architectural and Horticultural Insight, and Ideas from Famous Gardens in Major Civilizations

One of the most significant books on landscaping!

This is one of the most comprehensive books on planting design. It fills in the blanks of the field and introduces poetry, painting, and symbolism into planting design. It covers in detail the two major systems of planting design: formal planting design and naturalistic planting design. It has numerous line drawings and photos to illustrate the planting design concepts and principles. Through in-depth discussions of historical precedents and practical case studies, it uncovers the fundamental design principles and concepts, as well as the underpinning philosophy for planting design. It is an indispensable reference book for landscape architecture students, designers, architects, urban planners, and ordinary garden lovers.

What Others Are Saying about *Planting Design Illustrated* …

"I found this book to be absolutely fascinating. You will need to concentrate while reading it, but the effort will be well worth your time."
—Bobbie Schwartz, former president of APLD (Association of Professional Landscape Designers) and author of *The Design Puzzle: Putting the Pieces Together*.

"This is a book that you have to read, and it is more than well worth your time. Gang Chen takes you well beyond what you will learn in other books about basic principles like color, texture, and mass."
—Jane Berger, editor & publisher of gardendesignonline

"As a longtime consumer of gardening books, I am impressed with Gang Chen's inclusion of new information on planting design theory for Chinese and Japanese gardens. Many gardening books discuss the beauty of Japanese gardens, and a few discuss the unique charms of Chinese gardens, but this one explains how Japanese and Chinese history, as well as geography and artistic traditions, bear on the development of each country's style. The material on traditional Western garden planting is thorough and inspiring, too. *Planting Design Illustrated* definitely rewards repeated reading and study. Any garden designer will read it with profit."
—Jan Whitner, editor of the *Washington Park Arboretum Bulletin*

"Enhanced with an annotated bibliography and informative appendices, *Planting Design Illustrated* offers an especially "reader friendly" and practical guide that makes it a very strongly recommended addition to personal, professional, academic, and community library gardening & landscaping reference collection and supplemental reading list."
—Midwest Book Review

"Where to start? *Planting Design Illustrated* is, above all, fascinating and refreshing! Not something the lay reader encounters every day, the book presents an unlikely topic in an easily digestible, easy-to-follow way. It is superbly organized with a comprehensive table of contents, bibliography, and appendices. The writing, though expertly informative, maintains its accessibility throughout and is a joy to read. The detailed and beautiful illustrations expanding on the concepts presented were my favorite portion. One of the finest books I've encountered in this contest in the past 5 years."
—Writer's Digest 16th Annual International Self-Published Book Awards Judge's Commentary

"The work in my view has incredible application to planting design generally and a system approach to what is a very difficult subject to teach, at least in my experience. Also featured is a very beautiful philosophy of garden design principles bordering poetry. It's my strong conviction that this work needs to see the light of day by being published for the use of professionals, students & garden enthusiasts."
—Donald C. Brinkerhoff, FASLA, chairman and CEO of Lifescapes International, Inc.

Index

Made in the USA
San Bernardino, CA
22 May 2016